MINIATURE THERMOPLASTICS SCULPTURE

A Challenging Art Form

Some other publications by the author:

The Life History and Cytology of the Foraminiferan *Allogromia laticollaris.*
Univ. of California Publications in Zoology. vol. 61. 1955

Biological Observations on the Foraminifer *Spiroloculina hyalina* Schulze.
Univ. of California Publications in Zoology. vol. 72. 1964

Observations on the Biology of the Protozoan *Gromia oviformis. Univ. of
California Publications in Zoology.* vol. 100. 1972

A Technical Manual for the Biologist. Boxwood Press. 1993

Musical Punstruments. Boxwood Press. 1994

Cats and Yammerings. Boxwood Press. 1994

The Miniature Harmonica. Boxwood Press. *1995*

Musical Compositions (for the piano) published by the composer

Album Maritime
Berkeley Sketches
Concert Dances
Lyric Suite
Point du Jour
Romantic Suite
Waltz Suite
Pieces for the Piano
Réflexion
Christmas Album (piano and voice)

MINIATURE THERMOPLASTICS SCULPTURE

A CHALLENGING ART FORM

BY

ZACH M. ARNOLD

Illustrations by the author

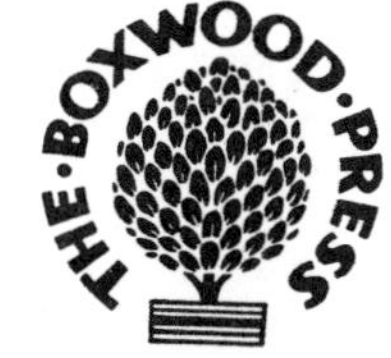

Pacific Grove, California

Distributed by:

The Boxwood Press
183 Ocean View Blvd.
Pacific Grove, CA 93950

Phone: 408—375-9110 * *Fax:* 408—375-0430

ISBN: 0- 940168-39-1

Printed in USA

PREFACE

This book is the outgrowth of a confluence of my diversified professional and non-professional interests and experiences, facilitated and sweetened by physical adversity and supplemented by sublimated insomnia and ailurophilic somnioclasy.

My basic interest in plastics developed in response to my need to gain sufficient technical mastery of the various processes of working with these materials to be able to apply them effectively in biological research on microscopic marine organisms. A natural counterpart to this need was that of using the techniques in various other professional activities as a scientist and teacher. The technical mastery essential to successful application within these two closely integrated spheres of activity was achieved through study, experimentation, and practice, largely in a home laboratory and workshop, basically through trial and error, and all financially on a shoestring.

My lifelong love for and addiction to tinkering and gadgeteering made tools and a home workshop essential to my well-being, and these, in turn, fostered and made feasible the endless experimentation engendered by a desire to invent and create things. It is difficult to say precisely when the fascination with miniaturization arose, but certainly it was there when, in my sophomore year at university, I saw through a microscope my first foraminiferan shell and fell immediately and irrevocably in love with the minute creatures that produce such natural marvels of miniature sculpture and masonry. The fascination with miniature sculpture had its roots in the vertebrate preparation laboratory of the Musuem of Paleontology (U.C., Berkeley) with the whimsical clay miniatures left about on the work tables by Philip Goerl and Wann Langston, and with Paul Lawson, who, while on loan from the Adelaide (Australia) museum, gave early encouragement to my budding interest in plastics technology, interests that continued to develop as an adjunct to my teaching and research at the university.

Our two daughters, Susan and Jane, introduced me, as a consequence of a summer course they took at the Richmond Art Center, to ceramic clay and ceramic sculpture, which stood me in good stead for many years of biological model making. From this to thermoplastics sculpture was an easy and natural transition in the light of previous experience with plastics technology.

When coronary problems made early retirement from the university advisable, the opportunity remained of pursuing, in a limited way and at a greatly reduced pace, the work with plastics, with gadgeteering, and with miniaturization. The step from making models of foraminifera as teaching aids to the construction of ship models, toys and novel musical instruments (*punstruments*, based on musical puns) was a simple one, a natural outgrowth of various technical aspects of my earlier professional activities, these later ones a satisfying suite of replacements for the earlier challenges, the shift toward an emphasis on the more artistic aspects an added challenge and source of satisfaction. (When, during various recuperative periods, the need to spend time in bed increased, means were devised for pursuing these sculptural interests, principally with clay; but with improved health, attention could soon be focussed on thermoplastics and the development of the tools and techniques for executing pieces in a material more easily worked in small scale than clay and more durable than wax.)

Insomnia, increasing with age, came as an unexpected blessing, another of the sweet uses of adversity, providing untold hours, readily salvageable from the treasures normally squandered in luxurious slumber and used now instead in reading art history, there encountering, belatedly, such delights as the Limbourg brothers' *Les Très Riches Heures du duc de Berry* and Giotto's dejected monk in *The Vision of the Burning Chariot* in which latter one sees the grandeur of simple line and form later so exquisitely captured by Brancusi in his *Mademoiselle Pogany* and by Canova in his *Mansuetudine.*

Zach M. Arnold

El Sobrante, Ca.
Spring, 1996

ACKNOWLEDGEMENTS

I am indebted to many craftsmen, technicians, scientists, and artists for the encouragement and inspiration essential to the develment of a deep-seated interest in and respect for true craftsmanship, the plastic arts generally, and miniature thermoplastics sculpture in particular. In my early teens Nathaniel Nungezer introduced me to fine woods and the joys of woodworking, an interest furthered by Mr. Stanford, an English cabinet maker, and Miss Jean Pick, an occupational therapist and physiotherapist (shortly to become my wife) in the occupational therapy section of the 16th Station Hospital (U.S. Army) in Bromsgrove, England during World War II. In the Museum of Paleontology (and as mentioned in the Preface) the handiwork of Philip Goerl, Paul Lawson and Wann Langston intrigued me. Len Vigus, Don Helmgren, and Richard Bugbee, mechanicians in the workshops of the Geology, Zoology, and Psychology departments, opened several doors to the banausic side of life, while Dr. G. Dallas Hanna, Professor John Gullberg, and Dr. E.H. Myers were, for me, paragons in the application of technological mastery (particularly in the area of microtechniques) to the pursuit of scientific investigation and to the science of pedagogical museology. Engineers John Corl, Joseph Countrywood, Roger Maineri, Kenneth Hill, and Dan O'Donnell have done much to encourage and assist me in my amateurish efforts with technical problems for which I lacked the background, but an exposure to pyrognostics in undergraduate mineralogy under the tutelage of Professor J. G. Lester pre-adapted me for thermoplastics experimentation. My ship-modelling friends, particularly Harry Hall and George Karitianos, have been a constant inspiration toward excellence in creative artistic miniaturization, and Ms. Mary Taylor has suggested improvements in the illustrations. Carol Anderson has given valuable assistance in proofreading.

My wife, Jean, nurtured (largely by her father) in the tradition of the Bauhaus, William Morris, and various arts-and-crafts movements in Europe and Britain, and later herself trained in the Dorset School of Occupational Therapy at Barnsley Hall (Bromsgrove, England), has done much to provide in our home the milieu and matrix in which the development of craftsmanship and artistic expression is encouraged and abetted. Generously abrogating any extradotal rights she might have had to the exclusive use of her paraphernal art-history library, she has made it freely available whenever I needed it. I monopolized a two-foot stack of these books for months on end during the preparation of this book, and not a complaint from her.

Zach M. Arnold

WARNING AND DISCLAIMER

Anyone planning to work with thermoplastics should early be made aware of the potential hazards involved. The fumes from heated or molten plastics can be harmful, so effective steps should be taken at the outset to avoid breathing them. Remove them from the work area by means of an adequate exhaust system. The risk of fire is also a very real one, so particular care should be exercised whenever heat is applied. Avoid excess heat, keep the work area clean (no inflammable material lying about), have a fire extinguisher close at hand, and keep a bucket of water and a blanket nearby for dowsing and quenching any accidental fire that may be started. Molten plastic, brought accidentally into contact with the skin, clings there tenaciously and can produce unpleasant burns!

If you plan to work with thermoplastics, take the time to go to a good library and read extensively on the hazards of working with these materials and of the steps to be taken to ensure a safe working environment with them.

Some of the procedures and tools described here have been developed specifically for my own use in thermoplastics sculpture, and some of the tools (the hot-melt glue gun, for example) are used in unconventional ways or with materials for which they were not designed by the manufacturer. This book is basically a record of the ways in which I have used them and found them to be satisfactory within the parameters of safety I consider adequate for myself. Anyone wishing to apply them in a similar way should first make certain that their use falls within the more rigid parameters set by proper authority after proper testing for compliance with existing safety regulations. The hazards of cutting certain plastics with power tools, for example, can be formidable unless proper precautions are taken to avoid having the cutting elements snatch the plastic from the operator's hand and throw it violently against some vulnerable part of the body (the eyes, for example). All the usual safety devices (face shield, eye guards, proper hold-downs and push-sticks, etc.) should be used.

Neither the author nor the publisher can be held responsible for injury or accident resulting from any action of the reader who attempts to execute any of the suggestions or ideas in this book without having proper qualifications to do so or without first having obtained the necessary basic instruction and guidance or assistance to do the job in accordance with recognized standards of safe design and safe workshop procedure.

Zach M. Arnold

CONTENTS

PART ONE 1
GENERALITIES

PART TWO 12
BASIC TOOLS AND PROCEDURES

PART THREE 40
ADVANCED TOOLS AND PROCEDURES

PART FOUR 109
IN PURSUIT OF NUMEN

PART FIVE 134
STORING, VIEWING, AND DISPLAYING MINIATURE SCULPTURAL PIECES

PART SIX 153
A PICTORIAL GALLERY
OF SCULPTURAL MINIATURES

xi

This book is dedicated
to
Susan and Jane

PYROGNOSTICS AND PYROGENICS

Mineralogical blowpipes are delightful tools to use,
For much you stand to learn from flames of different hues;
But how, indeed, can pyrognostics help the sculptor bold
Who's only worked in wood or clay or carved in stone so cold?

If such a one would look about with pyrogenic intentions
And couple these in a home workshop with simple banausic inventions,
He'd very soon find a red-hot loop of resistance wire
Is a find and dandy substitute for the trusty blowpipe's fire.

A squeamish pyrophobe, of course, will shun all smoke and flame
And look for other, safer ways of gaining world-wide fame,
But the pyrophile's heated loop and the lovely melt it makes
—If one but thinks in miniature— is what such sculpture takes.

Just as natural pyrogenesis produces lovely rocks
For sculptors' metamorphoses and resultant arty shocks,
So one's plastic, which flows with heat, transformed before one's eyes,
Becomes a form of grace and charm, its beauty a lovely surprise!

PART ONE
GENERALITIES

1

INTRODUCTION

THIS BOOK, on the creation of pleasing sculptural miniatures from thermoplastic scrap, is the result of a response to several challenges presented to the craftsperson or artist by the mere presence of such materials in our everyday life. First, there is the simple challenge of trying to create something tangible and satisfying—something worth keeping and even to be treasured—from materials considered by most to be worthless and simply thrown away. Then there is the challenge facing anyone with a social conscience to do something, however slight, to help reduce the demands man makes on the environment and its resources, to minimize the offensiveness of his mere presence on our overcrowded planet, and to help beautify or otherwise improve the environment for us all. Added to this is the technical challenge of learning to use certain thermoplastics effectively and of combining other materials with them in satisfyingly creative ways to produce objects of artistic merit through craftsmanship of a high order. This last challenge embodies those of using conventional tools and techniques in a masterful (though sometimes unconventional or novel) way, but it also includes the challenge of developing, making, and using new ones as needed to achieve one's goals. A further challenge, one that intensifies the demands on one's technical skills, is to operate at a miniature scale, though several advantages inherent in this approach accrue from it and help to ameliorate the strain inevitable from working in miniature.

The advantages of working at a miniature scale, however, seem far to outweigh the disadvantages. Additional demands are, without question, made on one's manual dexterity, and greater precision is required in most of one's movements, but perseverance and practice will work wonders in improving the control one has over those movements and the skill with which one uses one's hands and fingers, while lighting and magnifying lenses, properly selected, installed and used, should relieve the visual strain. Most other physical demands on the craftsperson are reduced along with the reduced scale of the work. Additionally, attention to the ambience and arrangement of the work station, the layout of the tools and material, and the positioning of the worker can improve the general work conditions and concurrently reduce nervous strain and fatigue. All of this response to the increased demands of miniaturization can lead to an overall increase in the efficiency of and the pleasure derived from one's creative efforts.

The more obvious advantages of working in miniature are: reduced capital outlay, i.e., reduced expenditure on tools and equipment (smaller tools and machines generally being less expensive than larger ones); reduced operating costs (supplies and materials used in lesser amounts); and reduced maintenance costs. Less electrical energy is required for the operation of small tools and machines, and, too, the craftsperson generally expends less human energy in operating them. At every stage in the operation, then, economies in time, money, and energy are effected through miniaturization. Moreover, less space is required for such small operations, not only less space in the workplace and for the operator's own needs, but less space and fewer demands generally on the extra-personal environment whence come the tools, machines, materials, supplies, and energy essential to the operation.

As one learns to work more carefully and with greater precision and dexterity, one developes an increased awareness of and sensivity to relationships and consequences, an increased temperance and modesty in one's demands on the environment, and an increased awareness of and appreciation for detail. Concomitant with all of this is increased self reliance, independence, individualism, ingenuity, determination, and perseverance. Moreover, the skills and discipline acquired in gaining mastery in one area of miniaturization may readily be directed toward others, opening undreamt-of vistas within the mind-bogglingly massive universe of the minute. The challenge of developing imaginative expression through a technically demanding medium and at a technically demanding scale is intensified by the demands for: (1) genuine craftsmanship of a high order; (2) worthy aesthetic and artistic standards; (3) moral standards that evoke the best both from the creator and from the viewer; (4) self sufficiency; and (5) economy in all categories and at all levels of operation.

While the personal satisfaction of working within these constraints to create objects of interest and beauty may be sufficient in itself, one should, as a social organism, also recognize the challenge—nay, the responsibility—of presenting one's work in the most favorable way possible to the public gaze. With this in mind, some suggestions are given later (Chapter 16) for displaying and viewing the miniature sculptural pieces.

THE ART OF PLASTIC WASTE
(A Recycling Art Form)

A molten blob of plastic waste,
Some twisted strands of wire,
Bits of scrap amassed in haste,
Imagination set afire!

While one can pull and twist the mass
And shape it to his dreams,
Another fails to see it pass,
Transformed in what it seems.

For one it's merely wasting time,
Squandering a bit of life;
For another 'tis a gift sublime,
Relief, so sweet, from strife.

You'll find that it's no trick at all
With plastic waste to sculpt a piece,
To make wee figures short or tall
And help your skill increase.

Make your pieces small and neat,
An inch or less in height;
All the dozens you complete
Make quite a charming sight.

They'll take up very little space,
No trick at all to pack away;
So easy, too, to find a place
To put them on display.

The tiny pieces catch the eye.
For you they meet the urge
(From scrap that others just decry)
To see some art emerge.

(cont. p. 4)

(cont. from p. 3)

But isn't this so true in life?
There's art around us ev'rywhere;
The world with art is truly rife,
We need but see to find it there.

There's art in how we turn a screw
Or how we drive a nail.
The artisan knows what to do
To use this art to best avail,

So cannot we, with plastic scrap,
Melt and form it to our taste,
Beauty latent there unwrap?
For us to find, it there was placed.

Just buckle down, get out the tools,
(You'll only need a few);
Be not concerned too much with rules,
You'll soon find out just what to do.

Behold a little shape appear
That's pleasing to the eye,
And then a further urge so clear
To have another try.

And so the days and weeks fly by;
The craze has taken hold!
The tougher the task, the harder you try:
Indeed, a way to joy untold!

THERE'S ART IN ALL THINGS,
IF MAN HAD EYES

There's art in all things, if man had eyes,
Which comes to us as no surprise
When once to it some thought is given
By those who long for art have striven.

Latent art in Max Ernst's *frottage,*
Shimmering art in a desert mirage.
There's art, some think, in an open wound,
And art, of sorts, in a maiden swooned.

Murmuring art in a rippling stream,
Somnolent art in a pleasant dream.
There's art, indeed, in cloud-filled skies,
And art in vats of aniline dyes.

Reticent art in a diffident pal,
Frivolous art in a flappery gal,
Merlinous art in a magician's wand,
Slithery art in a frog-filled pond.

It's often said that art you'll find
In the scheming plots of a devious mind,
But surely by far a nobler end
Is the lovely art of making a friend.

Some find art in maths and trig,
But some, no matter how hard they dig,
Can only find art in comic relief
From intense devotion to abstract belief.

Ah, yes, I strongly suspect
That if we but look we'll art detect
On ev'ry hand, in ev'ry nook
And not just in some picture book.

2

MATERIALS

MY UNTUTORED SEARCH for a material suited to the needs of my miniature sculptural urges began with wood at a time when my knowledge of tools and techniques was so rudimentary that I couldn't begin to achieve satisfactory results at the scale in which I wished to work. The difficulties and limitations of working with stone were long apparent from my early geological study of the textural and structural qualities of rocks and minerals. Moreover, as my requirements were for a material that could be worked not only in the true subtractive sense of classical sculpture but also in the additive sense of the modeller, both stone and wood were ruled out, as was ivory, a material difficult to obtain and immoral to use for anyone with an environmental conscience.

Early in my search, Allport's (1971) book on paper sculpture introduced me to that fascinating world and led tangentially to experiments with papier maché, but it soon became apparent that paper, in whatever form, was not suited to my needs. The challenge of clay sculpture was irresistible, as was that of ceramic sculpture, but, again, and after considerable experimentation, the technical difficulties of working at a miniature scale with clay in any form seemed insurmountable. At that stage, wax seemed to be the appropriate alternative and, in conjunction with the lost-wax or investment casting technique, the ideal solution, as indeed it has long proved to be with jewellers, dentists, industrial designers of many and varied hue, and miniaturists across a broad spectrum. By substituting polyester or epoxy resins for gold and silver in the final casting, the cost of the operation was greatly reduced and I managed to produce an impressively large suite of models for use as teaching aids at the desired scale, but the method still remained so time consuming as to be restrictive. The problem now was to find a way of working directly in a material that would become the final piece without the need of investing a delicate wax model and then reproducing it in another material through the casting process.

Wax, a natural thermoplastic, was a superb material for making the original model or pattern, so it seemed logical to seek a synthetic thermoplastic that would have the strength and durability lacking in wax but essential to the finished piece. After years of experimentation and experience in working with various materials, the ones best suited to my needs gradually became apparent. It is to the description of these and of the tools and techniques used in working them that this book is devoted.

The principal materials used in the various sculptural pieces described here are polyethylene and polypropylene, mostly the high density but also the low density types of the former, and all derived principally from such food containers as those in which yoghurt and margarine are packaged, although polyethylene from other household objects as buckets and basins is also useful, particularly when its color has a special appeal or a need for such color exists in making a given sculptural piece. Yoghurt and margarine containers form the bulk of the finished pieces described and illustrated in this book, their color and texture most resembling the ivory or stone that a sculptural classicist might usually prefer, but variety and interest are lent by plastics of different color, derived from a heterogeneous assortment of scrap cut from buckets, basins, children's toys, and all manner of plastics castoffs. In general, I prefer the pint or quart size of yoghurt container for most purposes, the plastic in these being thicker, although it is wise to have some thinner stock (from smaller containers) on hand as well for special needs. Most of the stock can best be kept in the form of thin strips (1-2 mm. wide and a few inches long), which are suitable for most of the demands of the sculpture attempted at miniature scale with minimal facilities. As one acquires skill and expands one's horizons, a more varied assortment of stock pieces, particularly extruded types, will be found useful. The small rectangular plates of plastic used as bag closures, can, for example, be a worthwhile source of colored flat stock.

In selecting scrap for working stock, the drawing or spinning properties of the plastic are most important considerations. It is well to test these by heating a small quantity with the electric-loop tool (to be described later) and attempting to pull it out into a string or thread, using a small needle or pin tool for this. Scrap that can be drawn out effectively in this way will not be too brittle for use in sculptural pieces. During the heating process the material should not char or discolor and its surface appearance after being drawn out should look much as it did before. With some experimentation it often is possible to improve the drawing or spinning properties of uncooperative stock by adding other substances to it, but for most purposes it seems generally wiser at the outset to select stocks having the desired properties rather than entering the more complicated realm of custom blending, the one obvious exception being in the use of thermoplastic glue, which blends well and easily and has, in my experience, given highly satisfactory results as an additive for a varied assortment of recalcitrant stocks in my collection.

Other properties that are to be considered as one builds up stocks of plastic are adhesiveness and cohesiveness (easily improved by the addition of thermoplastic glue), melting point, color, ease of forming by blowing, draping, hot-stamping, and other thermoforming processes.

Because many of the materials used in thermoplastics sculpture are potentially dangerous, particularly when heated, one should become familiar with the hazards and exercise proper caution when using them. The fumes from heated or molten plastics can be harmful to breathe, the plastics can burst into flame if heated too vigorously, and molten plastic can produce burns if allowed to come into contact with the skin. The miniaturist should become aware of these and other hazards and be properly respectful of them by taking all necessary safety precautions in the use of thermoplastics.

In order to select plastics efficiently from scrap materials, one should learn to interpret the identifying letters and numbers stamped on the bottom of food containers. The ones I have found most useful are the following:

> Low density polyethylene (LDPE)
> High density polyethylene (HDPE)
> Polypropylene (PP)
> Polyethylene terephthalate (PETE)

I have found it wise to avoid polystyrene (PS) and the vinyl plastics. A simple, quick test of the suitability of a given piece of unmarked scrap is to heat it carefully with the electric-loop tool described in the next chapter or a low flame from a Bunsen burner or propane torch to see if it melts cleanly (without discoloration, scorching, or charring) to form a thick, heavy, molten blob that can then be pulled out easily into a string or thread that is not excessively brittle when cooled. In a few instances, of course, one may be able to use charred material to advantage in achieving some particular effect, but as a general rule it is best to avoid such plastics, there being ample scope for artistic expression with the materials that melt easily and cleanly. Mixing a more cooperative material with one that seems less promising often can, however, give interesting results and broaden one's design horizons, the working properties of the composite often being adequately favorable. By mixing a mildly charred piece of plastic with one of favorable quality, for example, a lovely swirly brown can be produced and used to advantage.

An important component of one's sculptural stock is thermoplastic glue of the type used in glue guns, either the miniature or the regular ones, but the material is specially prepared for sculptural use by re-extruding it to form thin rods, threads and filaments which may then be used directly in their extruded form or subsequently combined or coextruded with other plastics in order to modify the working properties and appearance of the latter to one's needs and desires.

An assortment of waxes of various types, both natural and synthetic, can be of use, but waxes are used in only minor amounts and may be dispensed with altogether, although, when mixed with the plastics, the result has an appeal—both in working properties and in appearance—that is quite likely to capture one's fancy and even, if one is not careful, to displace polyethylene as the major component of one's sculptural program. Wax-working, as a sculptural technique and as the mainstay of many modelling and model-making efforts, is so well established and so extensively practiced that anyone planning to work effectively in miniature thermoplastics sculpture would be well advised to gain familiarity with its basics.While resisting the temptation to work exclusively in this delightful medium, and while keeping waxes in their proper subordinate role in our thermoplastics sculptural program, we still can use them effectively in many meaningful ways. The purist who restricts the materials to polyethylene and thermoplastic glue can produce a marvelous assortment of completed pieces and require nothing further for such achievement, but others may well wish to take advantage of the broader horizons opened through the use of other materials—particularly wood, metal and paper. Inasmuch as only minute quantities of any of these are required, they can well be (and should be) of good quality. The woods I use, all from my ship-modelling scrap bin, are principally boxwood, such fruitwoods as apple, pear, orange, and apricot, and such other hardwoods as California privet, beech, birch, and maple. Metals are represented principally by various fine-gauge wires (brass, copper, aluminum, and tungsten) and the spiral waste turned from aluminum and brass in the lathe. Such spiral scrap can add variety and interest to the sculptural pieces, as can similar waste produced from turning various synthetic plastics, both thermosetting and thermoplastic types. Paper of the type used in dinner napkins, nicely absorbent, lends itself to the production of interesting shapes when crumpled and dipped in molten wax, and single-ply drawing paper (Strathmore, for example) having a hard surface can be useful in a number of ways, as, for example, in making patterns and mockups for pieces or components to be developed later in plastic or as structural supports within the finished piece itself.

Fine wire and the plastic monofilaments (leader line particularly) used by fishermen can be of use in special applications when the delicate and highly heat-sensitive polyethylene threads and strings are not suitable. The strings used in making harps shown in Plate 6, Fig.h, for example, are most easily mounted if they are metal rather than polyethylene, because at such a small scale it is difficult to apply sufficient heat to fuse the string to the harp frame without at the same time melting the highly sensitive plastic-string material.

Various materials can be employed to provide support for miniature sculptures, although most pieces may be made without support. Often, however, the use of a support—either internal or external—can simplify the task or can enable one to expand one's design ambitions. Armatures of various types may be made from wire in its simple form, twisted, or built up by soldering. An axis, armature, or core may also be made of paper and subsequently encased by melting plastic over and around it. For this purpose, scrap drawing paper is ideal. One way of converting the paper into *papier maché* is by such rigorously hygenic but rather impersonal processes as cutting it up with scissors, soaking in water, and then blending with an electric blender. Another—equally hygenic and impersonal—is to use a belt sander (bench-mounted type) to convert the paper to a fine flour, which is then merely wetted and blended. When the excess water is removed and white glue added, forms may be built up which can then be encased in or coated with molten plastic to produce the final sculpture. A more intimate procedure—involving strictly personal projects in which no outsider is involved—is to chew the paper until it can be shaped as desired and used as the form to be coated with plastic. A wondrous array of marvelously imaginative forms can be produced in chewed paper—a craft form having a vast potential, and about as simple and cheap as any imaginable. (Such chewed-paper elements should, of course, be well dried before being coated with plastics.) The miniaturist should also consider using pieces formed in paper as a base over which to develope finished pieces in thermoplastics, using a cameo-making procedure like that described on p. 107.

The sculptor who prefers to work within the range of colors encountered in salvaged commercially-produced plastics may find these adequate to his or her needs, and by laying in a stock of scrap materials (principally polyethylene) encompassing a reasonably broad range may well find adequate scope, but if a particular color is not readily available in the scrap palette, it can be added at any one of three different stages in the sculptural process, namely:

   1. To the surface of the plastic before heating and melting.
   2. To the molten plastic (including the coextrusion of plastic stocks
      of different colors).
   3. To the surface of plastic not destined for further heating or
      melting.

The choice of stage is determined largely by the final effect one desires, each affording its own possibilities, but each with its own inherent limitations. Thus, if the pigment is applied to a plastic piece which is then heated sufficiently to be melted and drawn or otherwise deformed, the pigment will reflect the stresses applied to the plastic, their expression of these itself imparting interesting and characteristic features to the finished piece after cooling. Similarly, if the pigment is added to molten plastic, patterns reflecting the ensuing amount of stirring and mixing will be developed and expressed in the cooled piece or component. By contrast, color applied to the finished piece obeys essentially the same physical laws and aesthetics that govern any miniature painting, or, more properly—since the piece is three-dimensional—any painted sculptural piece. The more imaginative artist or craftperson would wisely be equipped to use a range of colorants and be prepared to apply them at whichever of the three stages gives the desired results. Be warned, however, that if you elect to use color in your sculpture you may quickly find yourself attempting to produce miniature paintings rather than mere painted sculptural pieces, the step from one to the other being a natural and easy one when the electric-loop tool of the miniature thermoplastics sculptor is substituted for the artist's brush, and thermoplastic materials are substituted for the artist's watercolors, oils, or acrylics.

Aesthetically speaking, an important advantage of resorting to such thermoplastics as high-density polyethylene in miniature sculpture is that one can easily achieve the elegant effect of ivory without in any way becoming involved in the legal or moral issues surrounding the use of the precious natural substance. If the plastic material is carefully chosen for its color and cleverly worked by a skilled craftsman, it will be difficult indeed to detect the simulation by mere visual examination.

PART TWO

BASIC TOOLS AND PROCEDURES

3

BASIC TOOLS

MOST OF the homemade tools described here were designed and constructed specifically for use in work on miniatures of one type or another: microscopic marine animals, biological models of various other types, musical *punstruments* (novel musical instruments based on musical puns), toys, and ship models, but all have been selected for illustration and description here because of their usefulness in making sculptural miniatures. The construction of most of them should be well within the scope of a home craftsperson whose workshop contains the basic hand and power tools. Some of the tools or ones that are roughly equivalent may be purchased ready made, as, for example, a fume hood, a jeweller's tool for making wax-patterns, and various work holders or positioners, but the ones described here are relatively easy to make, are certainly less expensive than their commercially-produced counterpart, and, being custom-tailored for use in this particular type of sculptural activity and for the space available, are probably more efficient for the job.

The work space

An attractive feature of miniature thermoplastics sculpture is that it requires little space. An efficient operation can initially be conducted on as little as two or three square feet of table surface if one plans the facility with care. In addition to the space occupied by an electric fan or by the fume hood in which most of the work at the beginner's level can be done, one needs spreading and storage space for the materials used in the work and for the finished pieces. As one's interest grows and activities expand, more space is required, of course, but in comparison to many relatively complex pursuits, this one makes but modest demands on space. A rolling chair is an advantage at the work place, as is a lamp on a swivelling arm, although a light built into the fume hood is a principal source of illumination for most of the work and may, indeed, be the only one required for satisfactory work. If one uses an electric fan instead of a fume hood for the removal of fumes, then a desk lamp of some sort is all that is required.

Fume hood

As a safeguard for one's health, primary consideration must be given to the effective removal of noxious fumes from the work area and the work room in which the plastics are heated or melted. The simple expedient of using an electric fan to move the air across the work bench will accomplish this effectively if open windows provide proper circulation, a rudimentary test of this system's adequacy being that no fumes can be smelled when work is being done on the plastics. To be on the safe side, however, in case one prefers not to trust an olfactory system that might be faulty, it is wiser to work inside a small fume hood purchased for the task or made in the home workshop. Various types of hood are available commercially, but these are usually quite expensive, require more space than is actually needed by the miniaturist, and lack some of the special features described in the homemade model illustrated here. A critical component of the hood is a blower (squirrel-cage type or equivalent) sufficiently powerful to move fumes and smoke rapidly and completely from the hood. The one I have used for years is powered by a 115-volt, 60-cycle, 0.21 amp. motor (Type 21, made by Fasco Ind., Inc.). The exhaust tube for the blower consists of a vacuum-cleaner hose, attached by a leak-proof joint to the blower. A simple way of expelling the air from the hood through a nearby window is to pass the free end of the hose to a plywood or Masonite panel cut to fit snugly into the space left by opening the window six to eight inches. Care should be taken to ascertain that all fumes pass to the outside and do not seep back into the room.

The basic hood, built of 0.250" plywood, is shown in Fig.1 .Two circular openings cut in the lower front wall permit one easily to pass small tools and work pieces into the work area and to work in comfort and safety. If fumes can be detected, a more powerful blower should be used, and to be on the safe side one can take the further precaution of adding leak-proof sleeves to the hand holes. The sloping upper half of the front is closed by a 0.187" sheet of clear acrylic plastic, the illustrated one sliding in grooves from the right side of the hood and so fitted as to preclude the escape of any fumes from the work area. It is a wise precaution to seal all joints with wood glue as the hood is assembled and to patch any openings (as for the fluorescent light ballast, for example) with automobile body cement. Test the hood carefully for escaping fumes before it is put into use and again periodically during regular work periods to make certain that all joints are tight and leakproof. Maintain such diligence whenever the hood is in use.

The following accessories and fittings make the basic hood a more useful facility and are strongly recommended:

1. A frame atop the hood in which to place and secure the control box for the electric-loop tool. This should be so placed that the controls are readily accessible and that the tool can easily be inserted into the hood for use. One or more spring clamps to hold additional electric-loop tools can be mounted on the outside wall of the hood in a position from which they can easily be attached to the control unit and inserted into the hood for use when required.

2. A foot switch to control the blower and the electric-loop tool simultaneously. Such a switch, easily made by mounting a microswitch inside a homemade case (Fig. 3, a) insures the operation of the blower whenever the electric tool is in use, the sound of the blower itself a handy reminder that the tool is activated, as is the pilot light on the control box. If one's blower is not adequate for the job and if smoke from the plastic is still visible in the hood after the electric-loop tool is removed from the work piece, this should be cleared by keeping the foot switch in operation for a few seconds. If this is not adequate, a more powerful blower should be used.

3. A fluorescent light, mounted inside the top wall of the hood. A 9- or 11-inch tube should provide ample illumination for a hood measuring 10 (deep) by 16" (long) by 9" (high), as mine does. The tube itself should not be visible to the operator, the ballast should be mounted atop the hood (outside), and the light switch should be mounted at a point on the hood convenient to the operator (mine is outside the hood).

4. Two or three panels of posterboard, in colors that provide a suitable range of background contrasts for viewing one's work efficiently, should be prepared for attachment to the inside wall of the hood, behind the actual work area. My hood is painted white inside, this assuring efficient overall use of the available illumination, but often the contrast afforded by a strategically-placed colored panel makes it easier to see details of concern as one works on a piece. The panels should be easily removable and interchangeable, blobs of soft wax (like that used to package some types of cheese) excellent for holding them in place but less elegant than custom-designed clamps made from bits of scrap packing bands (steel).

5. A magnifying lens, mounted outside the hood on an adjustable arm so placed that the work area can be viewed critically, is of considerable value—particularly to those of us with ageing eyesight— in achieving and maintaining the standards essential to a challenging and satisfying accomplishment in miniature sculpture. For some months I happily used a stereoscopic loupe (mounted on the forehead) for the purpose, but a mounted five-inch lens attached above the viewing panel of the hood itself is much more convenient. In some cases, however, it is still necessary to supplement this with the loupe or on occasion even to swing the mounted lens to one side so that the entire work area can be viewed with the loupe alone. A removable cover for the mounted lens is desirable to prevent accidental scratching and to keep it dust free.

6. A cradle, or caddy, for the electric-loop tool. The cradle shown in Fig. 2, h has a heavy ball-and-socket base which gives it both stability and adjustability, enabling one to position the tool with precision, and conveniently freeing both hands to manipulate a work piece and accessory tools when necessary. The tool can be rested loosely in the cradle or clamped securely when this is advisable, as when it must be used in heating mandrels needed in shaping plastic stock to curvatures not easily achieved without them, both hands thus being freed for the operation. It is also useful (Fig. 27) when filling small molds with molten plastic and pressing the hot melt into the mold to assure accurate reproduction, a procedure requiring both hands if it is to be done efficiently. One method of making knobs for the upper section of the cradle is shown in Fig. 2, k: a machine screw embedded in epoxy cement cast in the salvaged cap from an empty toothpaste tube.

7. A simple spring clip should be mounted high on the back inner wall of the hood to hold the electric-loop tool when it is not in use. The clamp (Fig. 2, f) is easily made from a scrap piece of steel strapping from a packing crate, the shaping efficiently accomplished by using metal rods of appropriate diameter as mandrels (clamped in a vise) around which to form it (Fig. 2, a-e). This clip serves as a supplement to a caddy for the tool, the caddy for use in holding the tool in its activated (heated) condition or for intermittent holding (but ready for instant use) when the tool hand is required elsewhere, the clip for storage at the end of a work session.

8. Shelves inside the hood, to hold supplies and other materials used in making the sculptural pieces. Some space should be reserved inside the hood for caddies in which favored stocks of plastics and threads, strings, and rods of hot-melt glue can be kept close at hand.

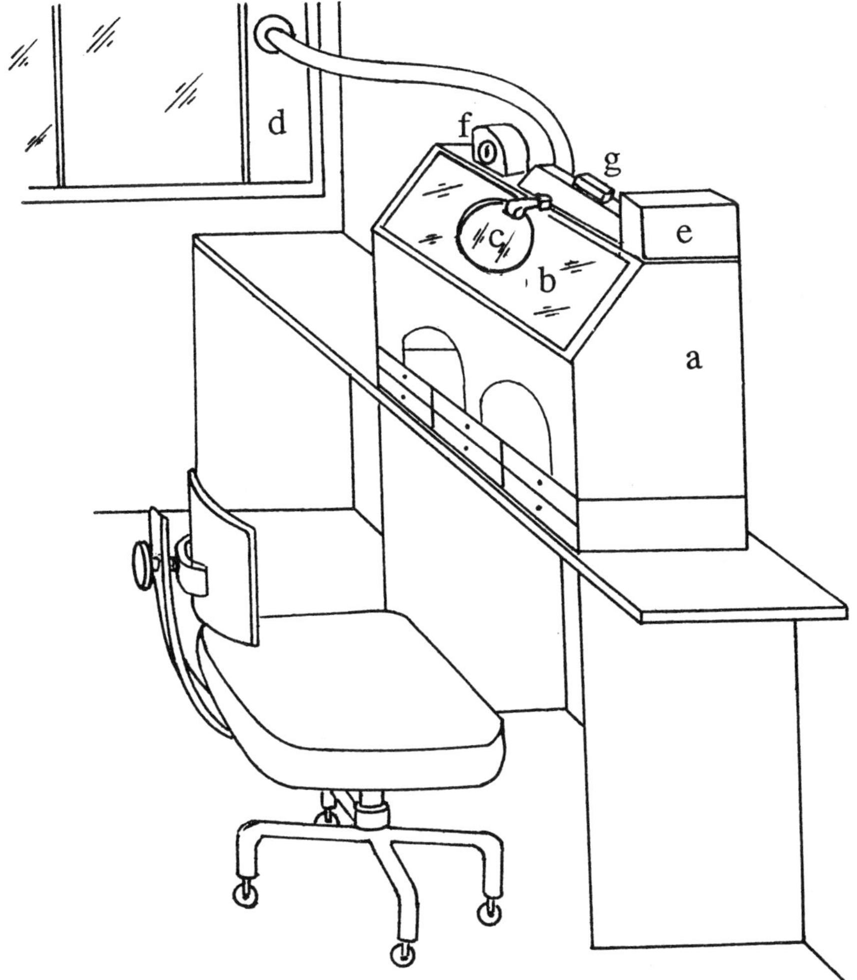

Fig. 1. Work station for miniature thermoplastics sculpture. a. Fume hood. b. Viewing window (plastic) for work area. c. Magnifying lens. d. Panel insert for holding exhaust outlet in space provided by partly opened window. e. Control unit for electric-loop tool (see Fig. 3). f. Control unit for miniature lathe/potter's wheel (see Fig. 24). g. Ballast and shield for fluorescent light. Fume hood dimensions: 18 x 14 x 10.5 inches.

9. Two tiers of small trays for other tools used in the sculptural activities are designed as part of the hood, fitted below its work area. These hold other small tools needed at various stages in the operation, keeping them conveniently at hand without cluttering the actual work area.

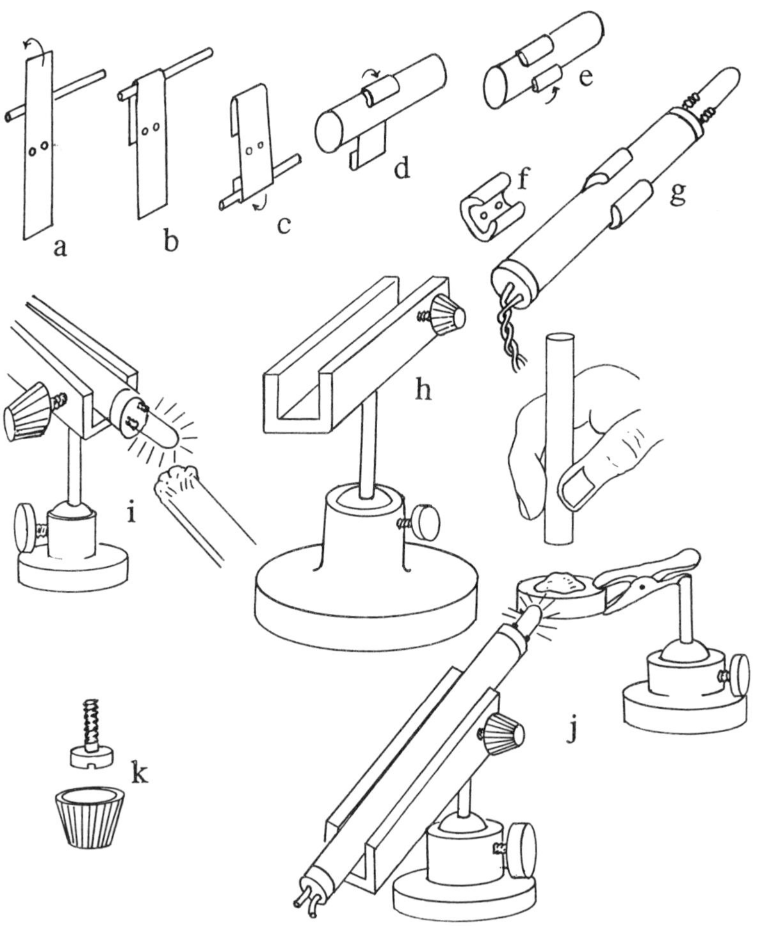

Fig. 2. Tool holder and positioner for electric-loop tool. a-g. Clip for holding electric-loop tool on inside wall of fume hood. a-e. Steps in forming spring clip. f. Finished clip. g. Electric-loop tool in its holder. h. Positioner for electric-loop tool. i. Positioner in use, holding the tool while melting plastic stock. j. Positioner holding electric-loop tool while heat-softening plastic in a mold prior to filling the mold with the aid of a tamping stick. k. Components of a homemade knob (toothpaste tube cap and machine screw) of the type shown on the tool positioner.

Electric-loop tool

The most important tool in the kit for making miniature sculptural pieces from thermoplastics is this one, merely a loop of nichrome resistance wire mounted in a convenient holder and operated by controls that permit one to select from a wide range of temperatures the one specifically suited to the particular needs at the moment. The size and shape of the loop can readily be altered to those most appropriate at the time, and one loop can easily and quickly be replaced by another, all at negligible cost. If one wishes a heat-tip smaller than that achievable by bending the loop upon itself, this can be obtained by inserting a pin or needle in a hole provided at the base of the loop and then wrapping the loop around it to concentrate the heat on this new working tip. An alternative procedure is to attach a pin or needle to the loop by means of a small alligator clip (Fig. 3, f) so that this tip can be used as a carving tool for working extremely fine detail. A serious problem here, however, is that some molten plastic adheres to the tip and must be removed regularly, this being accomplished by making a quick pass through a cold and uncritical part of the work piece before returning to the critical area. With a little experimentation one could, no doubt, produce special slip-on tips of various shapes, roughly paralleling the types supplied with jeweller's wax-pattern-working tools, but the crude system just described has served me well and, hopefully, will so serve the reader. I find it convenient to have two or three separate holders, each equipped with a loop, pin, or needle of a different size, the one of choice plugged into the control unit as needed. My own collection of holders is varied, ranging from discarded fountain-pen bodies to lengths of chrome-plated copper tubing of the type used in install-ing plumbing fixtures, this latter, though slightly larger in diameter than the plastic one from fountain pens, more durable and less likely to be softened or deformed by the heat during protracted use.

The essential features of the holder are shown in Fig. 3, b. The resistance wire is held in place by set screws (00-90) in the two brass terminals turned down as shown. The cooling fins are turned at the free end and a hole is drilled through the length of each terminal to receive the resistance wire. A wide shoulder is left at the base of the fin section to accommodate the set screw, and the remainder of each is turned to fit into the heat-resistant headpiece of the holder after being drilled (at its inner end) to receive the electric lead, which will be soldered in place within it. The head of my holder was turned from Transite (asbestos-containing cement) before asbestos be-

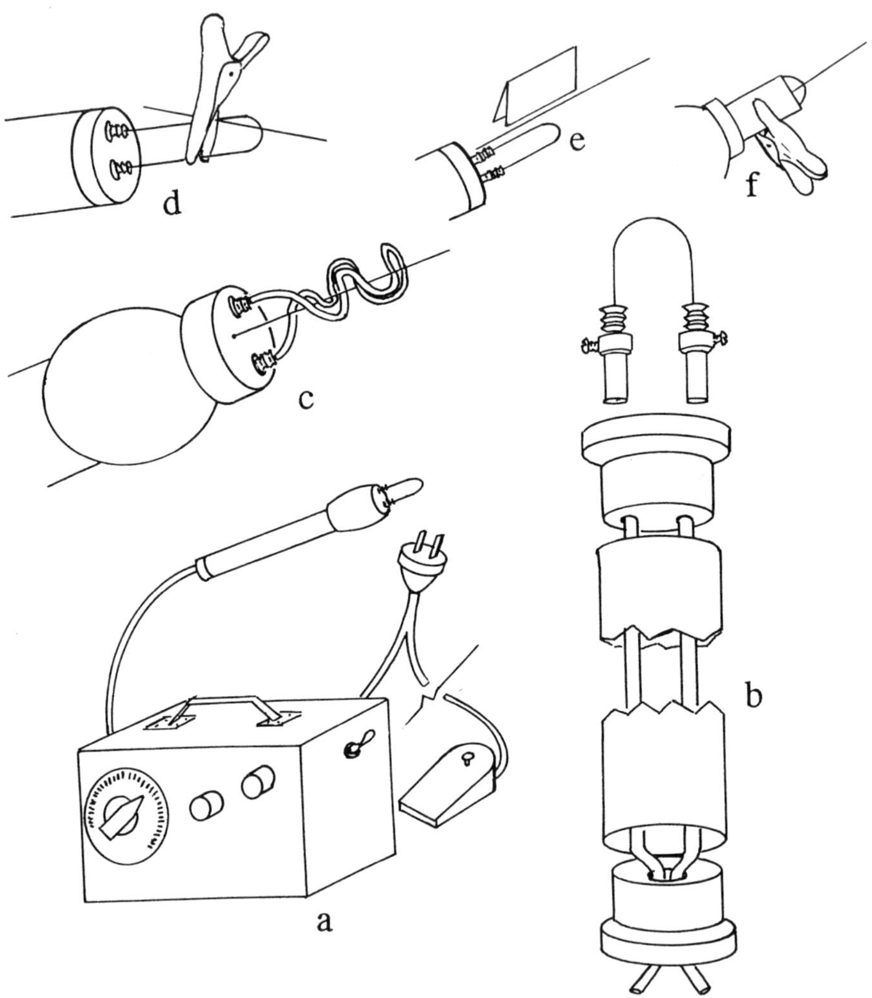

Fig. 3. The electric-loop tool. a. The tool, its controls box, and foot switch ready for use. b. Exploded view of the tool, showing its component parts. c. The head of the tool, showing the method of coiling the resistance loop around a straight pin, which thus becomes the working tip for precise heating and sculptural operations. [Note: a cork insulating sheath (ball) is shown in position around the tool shaft here but is an optional component.] d-f. Two alternate ways of affixing a pin to the electric loop: held against the loop with an alligator clamp (d), or sandwiched within a brass sheath (e) and held in place with an alligator clamp (f). Overall length of tool: 5.5 inches.

came anathema, but socially acceptable modern substitutes for this highly favored (and justly so) material from former generations should be relatively easy to find by consulting one's friendly hardware dealer or electrical supply source. The lower part of the head should be turned to seat securely inside the body tube and be either pinned or glued (heat-resistant cement) in place. The lead wires pass through a turned plastic plug (acrylic being adequate here, since the part does not become hot) at the opposite end of the body and are fitted with a plug for attachment to the control unit. The electrical controls for the unit consist of a small toroidally-wound transformer (Powerstat No. 10B or equivalent) operating through a 6-volt filament transformer. A toggle switch mounted on the side of the case is readily accessible to the operator when the unit is placed atop the hood. A fuse should be installed to protect the circuit, and a pilot light added to serve as a reminder to the operator that the current is on and ready for use. The unit is connected to a foot switch (Fig.3, a), as is the blower, which should automatically be engaged whenever the tool is activated, the sound of the blower being yet another safety feature designed to prevent the operator from intentionally or accidentally touching the heated tip even when it is not visibly red (with heat) but is still sufficiently hot to produce an unpleasant burn if contacted.

This electric-loop tool is essential in building up or removing plastic material from a sculptural mass to produce the details of the finished piece; it is the basic tool for the precise application of carefully-regulated heat required in making the material respond to the sculptor's wishes.

Work holders and positioners

Because of the diminutive size of the sculptural pieces, it is essential to have efficient holders and positioners for them and for any components upon which work must be done in producing the finished pieces. A critical feature of such tools is that they have a stable base that permits them to stand alone when in use, thus freeing both hands for the operation, one for manipulating the electric-loop tool, the other for manipulating the work piece and for using small tools (forceps, needles, etc.) in adding, removing, or changing the disposition of material. Various work positioners and holding or clamping devices may be purchased from suppliers catering to the needs of jewellers, the electronics industry, model makers, and workers in other crafts, but the ones illustrated here are easy to make in a home workshop and should meet most needs at a fraction of the cost of commercially-produced ones.

A case in point is the positioner shown in Fig. 4, a. This one, having a flexible stem that can be bent to about any position desired, has all the versatility one requires in most operations and can be assembled most inexpensively: an alligator clamp (from a radio/electronics store), a four-inch length of insulated copper wire (12-gauge), and a solid metal (brass) base drilled to accept the stripped (insulation removed) lower end of the wire and held in place with a knurled screw. The clamp is soldered to the upper, or free, end of the wire. The wire should be sufficiently flexible to permit one to move a workpiece to the proper position for shaping. Rotation is easily achieved by loosening the screw at the base or by rotating the entire tool, base and all. The base should be heavy enough to give stability to the work, and the positioner itself should be of a size convenient to the space available inside the hood.

The more elaborate positioner shown in Fig.4, b operates with a ball-and-socket mount, and the clamp at the working end is locked in position with a control screw that acts on the ball. The alligator clamp, soldered to a brass adaptor rod, is held securely in position by a setscrew which, when loosened, permits full rotation through 360 degrees, as does the screw at the base. A lathe is required to make the basal components of this positioner, but it has the advantage of being more compact, its low profile requiring less space within the hood, and being somewhat more rigid, although both have served me well and are used interchangeably. The ball-and-socket holder and the two knurled control screws of the second positioner provide a wide range of movement and rigid positioning, but to achieve the range of movement of which the first one is capable requires the use of three different (and inter-changeable) mounts with their clips: one with a vertical, or straight, arm; one with an arm bent at 90 degrees; and one bent at 45 degrees. This means that the mounts must be changed at various stages in the construction of some of the sculptural pieces, but such a procedure requires but seconds. In any case, all the alligator clips used for each of the positioners should be securely soldered to their mounting arm, and their teeth should be so modified (by grinding away the rearmost ones) as to insure a positive grip on the work piece. I usually find it most convenient and efficient to mount the work piece on a short length of plastic scrap or on a circular base (posterboard) that can be held securely by the modified jaws of the clamp.

Most of one's work will probably be done within the hood itself, and for this the two positioners just described will suffice, but should you decide to work with an electric fan instead of a hood, a positioner like that shown in

Fig. 4, d, requiring more space than the other two, can prove most useful, the basic features of its design permitting a satisfying range of movements. Constructed from salvaged parts of a superannuated microscope and other bits and pieces, the device has a heavy cast-iron base, which gives it stability. The entire superstructure can be rotated 360 degrees, the main stem can be raised or lowered through a range of 2 inches, the sleeve head can be rotated through approximately 170 degrees, and the arm can be rotated 360 degrees and slid through a range of 6 inches, each of these movements usable independently or in conjunction with one or more of the others, making it almost universal in its adjustment potential.

By removing the alligator clip from the arm, one can in its stead attach the small positioner shown in Fig. 4, e. When the precise assembly of small components of a sculptural piece is required, this clamp, made from acrylic plastic and three pins or needles fitted with acrylic knobs, can be most helpful. The pins or needles should be press-fitted into slightly undersized holes, so that they keep their position until aggressively moved by the operator. Components to be assembled are attached to the tips of the pins or needles by means of hot-melt glue (using the precise-gluing technique described elsewhere in this book).

Another positioner for use on the workbench (rather than in the hood) is shown in Fig.4, c. Assembled from a disk of acrylic plastic and a modified alligator clip (a brass or copper support strip soldered to the underside, all bent, drilled, and screwed to the base at approximately the angle shown), the positioner has numerous uses, not only in miniature sculpture but elsewhere (as a memo-holder, for example).

The positioner shown in Fig. 4, f, the smallest of the ones here considered, has proved useful in various operations, particularly in ancillary ones performed outside the hood, such as, for example, in holding miniature figures for painting or for building up bodies prior to adding the figures to a sculptural piece. Only 1.25 inches tall, the positioner, with its ball-and-socket mount for the clamp, is a versatile one and can easily be held in one hand, although it is stable enough to be used on the work bench, freeing both hands. The two different types of clamps may be used interchangeably here, the smaller one specifically designed to hold the fine wire used in miniature armatures, the larger (miniature alligator clip) to hold larger pieces or components. This small positioner was originally designed for use with a miniature clay-sculpturing facility intended for use in bed during a convalescent period, but it has served well subsequently in salvaging and using to

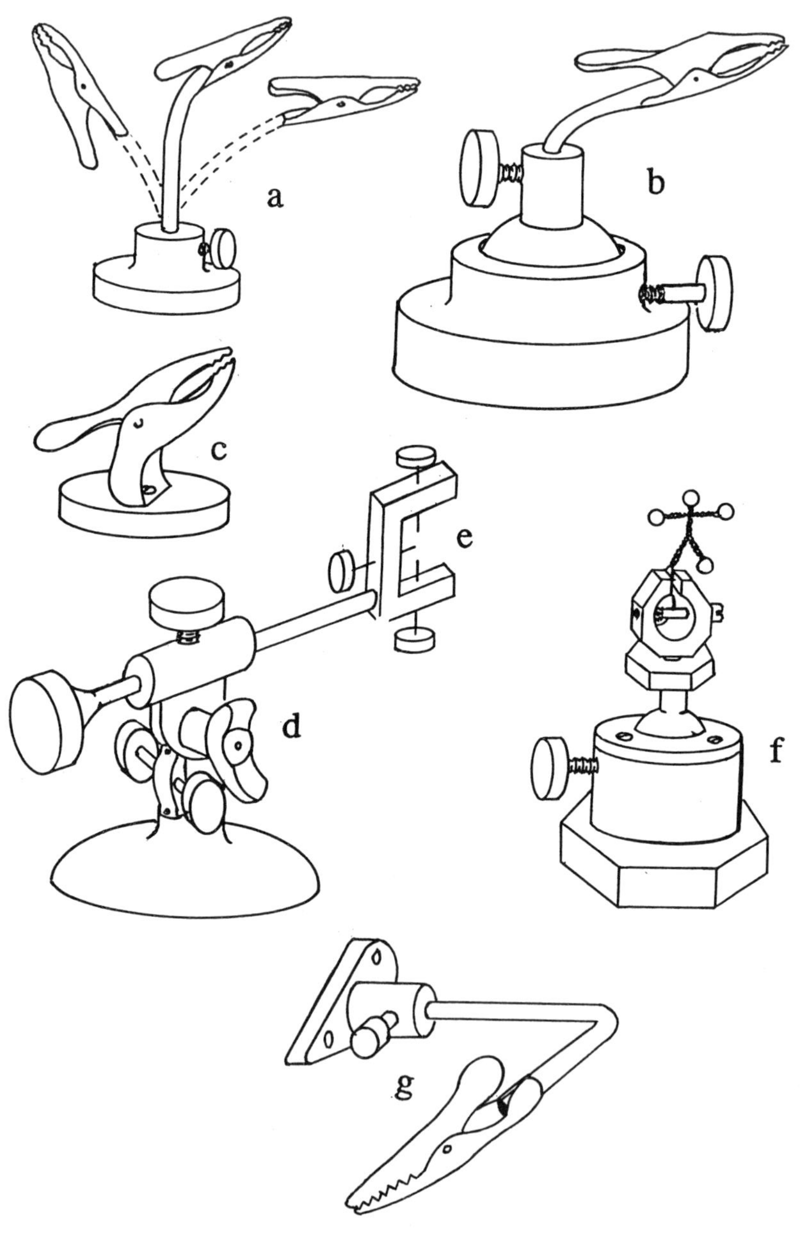

FIG. 4

good advantage those periods of nocturnal wakefulness that so frequently illumine increasing age (How sweet, the uses of adversity!).

The small copy stand shown in Fig. 4, g, used in holding an illustration of a sculptural piece to be copied during practice sessions, is merely an alligator clamp soldered to a short length of coated electric wire, the unit held (by means of a thumb screw) in a simple metal base attached to an outside wall of the fume hood at a spot convenient to the operator.

Miscellaneous small tools

A sturdy pair of scissors is an essential tool. With scissors, one can efficiently salvage the plastic from yoghurt and margarine containers and keep on hand a good working stock.

Small caddies (Fig. 5, a, b) in which to store one's stock, including the strings and threads of hot-melt glue specially extruded for the purpose, are a great convenience and are easily made from such household discards as frozen-juice cans and other small cardboard packaging containers.

A most useful pick-up and manipulatory tool (Fig. 5, l) is a short length of swab stick or dowelling with a blob of wax wrapped around what will be the working tip. The wax should be reasonably soft and easily shaped with the fingers. (The red wax used in packaging cheese is excellent for the purpose.) By shaping the tip appropriately, this simple tool is most helpful in picking up or manipulating and holding small work pieces and component parts of sculptural works. It becomes a versatile work positioner, too, if the free end is thrust into a larger blob of wax affixed to the floor of the fume

Fig. 4. Work positioners. a. Positioner made with an alligator clip, a length of flexible wire, and a heavy metal base. Overall height: 6 inches. b. Positioner made with a heavy swivel base (ball and socket) and an alligator clip fitted with a rigid shaft. A supplementary set of interchangeable clips (not shown) with shafts bent at various angles or straight enables one to position work at a convenient angle and height. Overall height: 4 inches. c. Simple positioner, a modified alligator clip and plastic (acrylic) base. Overall height: 2 inches. d. Versatile positioner made from parts of a superannuated microscope. Height: 6 inches. length: 8.5 inches. Shown here fitted with a small removable positioner (e) used in positioning and holding small components that are to be assembled (glued or heat-welded) in developing finished sculptural pieces. The small positioner is 1.25 inches high. f. Miniature positioner (1.25" high), particularly useful when working under a low-power microscope. g. Positioner that will be fitted to the outside of the fume hood to hold an illustration being copied (max. length: 9 inches).

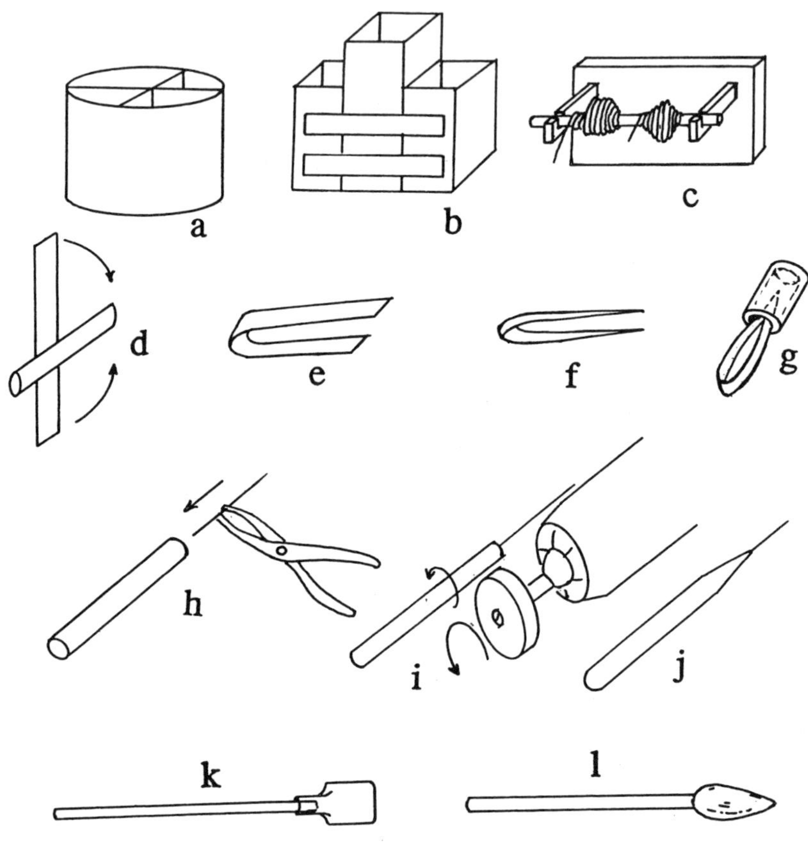

Fig. 5. Miscellaneous small tools. a-b. Caddies for extruded plastic rod stock. One (a) is made from a frozen-juice container cut to suitable height and fitted with cardboard partitions, the other (b) from toothpaste cartons, cut to suitable length and taped together. c. Reel for extruded hot-glue threads. d-g. Making simple tweezers: d. Bending scrap spring strapping over a mandrel. e. Bent blank ready for shaping (by grinding). f. Shaped tweezers, the tips ground. g. Polyethylene sheath for the sharpened tweezers. h-j. Making miniature needle tools: h. Pressing needle (one of the smallest of entomological pins) into a swab stick or fine dowelling. i. Tapering the head of the swab stick. j. Finished tool (overall length: 2.5 inches). k. Miniature homemade spatula (molded polyethylene blade fitted to swab stick). Overall length: 3 inches. l. Wax-tipped tool for picking up and manipulating small objects; overall length: 3 inches.

hood, the working tip then used as the surface to which the work piece is pressed for holding.

The remaining essential tools are a small pair of forceps and a needle mounted in a wooden or plastic handle. With the forceps, one can manipulate the work piece or component parts efficiently and with less risk of damaging them if one uses fingers alone, and with the needle one can then manipulate a molten mass of plastic in various ways while adding to or removing plastic from it with the electric-loop tool. The needle, used skillfully in conjuction with the latter, permits one to add such detail as facial features, which are difficult, if not impossible, to add with the loop alone; using the needle as a separate tool also in many cases eliminates the need to attach a needle tip to the loop tool.

Both forceps and mounted needles may be purchased, but adequate ones can be made in the home workshop quite easily. The forceps (Fig. 5, d-g)) are made from pieces of steel binding strap salvaged from a packing crate. A 3-inch length of half-inch strap is bent loosely in the middle; the ends are brought together and tapered on a bench grinder (dunking frequently in water to avoid drawing the temper) until the tips occlude sufficiently well for one easily to pick up and manipulate the smallest bits and pieces used in the sculptural effort. The needle tool is made simply by press-fitting a small needle (or decapitated pin) into a predrilled hole in the end of a short length of pretapered wooden dowelling. A particularly delicate needle tool can be made with one of the finest of entomological pins pressed into the end of an applicator stick, the end then tapered on a fine grinding wheel (Fig. 5, h-j). An assortment of needle tools of different sizes is a useful addition to the tool kit.

Needless to say, and in spite of the noblest intentions toward self sufficiency the craftsperson might entertain, one quickly finds as one proceeds with miniature activities that the investment in a good selection of professionally produced miniaturist tools, such as a pair of jeweller's forceps, end cutters, Swiss pattern files, etc., is a wise and worthwhile one.

A surprisingly useful tool for performing a variety of miniaturistic chores is the miniature spatula shown in Fig. 26, k, a simple polyethylene blade thermoformed (in a heated two-piece aluminum mold) from small-diameter tubing and fitted to a short (3") applicator stick.

4

BASIC PROCEDURES

Preliminary considerations

WITH THE BASIC TOOLS and materials already considered, a skilled worker can produce many of the pieces illustrated elsewhere in the book, and an imaginative one could, undoubtedly, produce an array as satisfying as many professional sculptors do in their work at larger scale, but the selection of procedures available to one equipped with only the basic tools is considerably less than it will be when the additional tools to be described later are available. If one attempts the additional procedures, however, one may initially feel that the demands on one's time and patience increase commensurately and the accompanying frustrations seem to become immeasurably more burdensome, but such is growth and development. More often than not, one finds that while some of the more interesting shapes and forms can indeed be produced through the use of the basic tools and procedures alone, the task of doing so can be made easier and less demanding by having at hand some of the addition tools and the mastery of the more elaborate procedures. It is wise, however, to gain a degree of competence with the basic ones before moving on the the others.

The basic sculptural and modelling procedures of removing or adding material may readily and quickly be accomplished, for example, on a human figure measuring only 10 mm. in height through the use of the electric-loop tool and a mounted needle (the latter to facilitate removal of the excess plastic when necessary), but by the mere addition of a small glue gun (for use in extruding rod, string and thread stock) and a tool for twisting wire armatures, figures—both human and animal, to say nothing of botanical and other forms—can be produced in a fraction of the time. Such a complex setting as that shown in Fig. 15,g will be difficult to produce with the electric-loop tool and needle alone, but with the simple additional facilities required in the *pulled-adhesion* technique described on pp. 57-61 it can be formed in less than a minute, the point here being that it is well worth while to enhance one facilities beyond the basics at an early stage in one's miniature sculptural efforts. If one has the patience and skill and is prepared to invest the necessary time, one can, indeed, achieve exquisite results with the simplest of tools and procedures, but the modest additional investment that

28

allows one to move into the more advanced procedures will pay handsome dividends, quickly enabling one to expand one's horizons and paving the way to undreamt-of accomplishment.

As an example of procedures one can employ successfully with only the basic tools, though not with the ease with which it could be done using the additional ones to be described later, consider the processes of drawing, pulling, spinning, and otherwise shaping and distorting a relatively small molten mass with forceps and mounted needle alone and with heat only from the electric-loop tool. These processes can be executed with such simple tools and, indeed, one may well find the results so satisfying that the additional ones are not necessary, but they should be considered at an early stage. The natural flow of the molten plastic under the strain and stress of the simplest tool action produces forms of pleasing shape as the flow developes and is then allowed to set. If, however, one chooses to work at a larger scale and use a bunsen burner as a heat source, an easy and effective way of producing fine threads of plastic is to melt one end of a strip of stock material to form a relatively large molten mass which is then allowed to drip and fall a distance of several inches to a foot or more onto a metal plate (Fig. 15, f). As the material falls from the stock piece, a fine thread is spun between the two and then further spun into greatly extended lengths by continued pulling (with tweezers or pliers) as long as the mass on the plate remains fluid. The greater distance the drop falls, the longer and finer the thread initially produced. As in any pulling, spinning, or drawing operation, the plastic selected for the purpose must itself have suitable properties, but a little experimentation soon enables one to select wisely in building up stock supplies. [A word of caution: *Melting the plastic is hazardous and should be done with great care to avoid setting it alight and starting a fire. Molten plastic causes severe burns if it contacts the skin, and the fumes can be dangerous. Keep a fire extinguisher close at hand and a fan to carry away fumes. Work in a clean, well-ventilated and uncluttered area—no flammable dust or trash nearby. Have a damp cloth and a bucket of water at hand for smothering any errant flame not sufficiently serious to warrant use of the fire extinguisher. (Throw a flaming piece of plastic into the bucket of water.)*]

As will be discussed subsequently, plastic thread may also be spun out of the molten extrudate from a miniature glue gun, but the method just described can be equally effective and is both simpler and faster, since the molten mass acts as a reservoir for spinning and can be made much larger than the one formed at the nozzle of a glue gun.

Michelangelo liked to distinguish between the *subtractive* or *carving* processes of "true" sculpture and the *additive* ones of modelling, but, fortunately for the miniature thermoplastics sculptor such distinctions blur almost irretrievably, because with these plastic materials one readily and naturally combines the two processes as one works, using the electric-loop tool to accomplish both. As a matter of fact, because of one of the properties of the material, namely the facility with which it can be drawn or pulled out to great lengths—a property found neither in the stone or wood of the classical sculptor nor in the clay of the modeller—a third category, one intermediate between sculpture and modelling, emerges, in a sense a blend of the two, in that the material is first pulled away from the main mass (analogous to *carving*) but, instead of being discarded as it would be in true sculpture, it frequently is returned and used to build up the final composition, i.e., *modelled* into the final form of the finished piece. Yet another feature of the thermoplastics, their durability and toughness, means that the sculptor's *maquette*, or model, is the final piece, not requiring transformation into a bronze or other type of casting, although it can, of course, be reproduced in metal (or in other plastic materials) by the lost-wax process if desired.

Preparing plastic stock for use

The plastic of yoghurt and margarine containers can merely be cut into thin strips with scissors or a paper cutter and these used directly as a welder uses welding rods to build up metal sculpture, or, as will be discussed in Chapter 8, they can be melted down and extruded in the form of rods, strings, or threads and, if one wishes to make the supplementary nozzles for the extruder needed in accomplishing such sophisticated tasks, as forms duplicating those normally produced from steel or aluminum in industry, namely, bar, angle, I-beam and U-beam stock. By using other procedures one can stamp or punch out special shapes, geometrical forms being especially easy to produce, but with simple molds one can replicate components in wide variety; with other plastics-forming techniques, such as vacuum-forming, blow-forming and blow-molding, yoke-plug molding, and forming with matched molds, one can add greatly (and in highly imaginative ways) to the variety and diversity of one's sculptural output.

Preparing a mount for the sculptural piece

Before work begins on a sculptural piece, a suitable base or plate should be provided on which it can be mounted. I prefer a base to which the piece will be permanently attached and thus become the base of the final mount. The size of the mounting plate is determined by the size intended for the finished piece, and the color of the plate should be chosen to contrast pleasingly with the piece itself. I find it convenient to keep on hand a stock of circular plates (Fig. 6. a-d) of various diameters (mostly 1.25 to 2 inches in diameter), these cut from poster board of good quality, its thickness 1.25 mm. The plates are readily cut with a gasket cutter of the type for use in a drill press, due care taken to operate it at slow speed and to keep one's fingers and hands out of the path of the whirling tool. To avoid having the tool snatch the board from one's grip, attach it to a piece of plywood that can be held securely and provide the leverage necessary for safety. *The operation is a potentially dangerous one, so great care must be exercised in performing it!*

The gasket cutter produces a central hole in each disk as it is cut, but this need not be offensive, as it usually is covered by the sculptural piece or can readily be filled (wood glue plus fine sawdust or wood flour) and painted to match the board itself. A stock of disks in two or three colors will fill most needs, the colors selected appropriate to the pieces one anticipates producing.

The plastic from which the sculpture is to be made must be attached securely to the mounting plate. This is best done by melting a quantity of hot-melt glue onto the attachment area of the plate and pressing the plastic into the molten mass, using the electric-loop tool to assure perfect adhesion between glue and plate and between the glue and the work piece. All must be pressed firmly together and held until the glue cools and effects a secure bond, usually a matter of seconds only. It is a simple matter with the electric loop to melt additional plastic for the work piece onto and into the glue to achieve perfect fusion and completely mask the glue, so that it is not discernible in the finished work. Once a satisfactory union has been achieved, work can begin on the sculpture itself.

In some cases, a supplementary base is required to enhance the sculptural design. This base, attached to the posterboard disk, can be made in various ways (to be described later), but if made of plastic can be produced in considerable variety both easily and inexpensively, using tools and procedures to be described latter. In addition to simple disks cut from plastic scrap with an arch punch (leather working), other types of punches, or with a hole cutter, such shapes as those in the figure may be produced by blow-forming, vacuum forming, yoke/plug molding, and forming with matched molds.

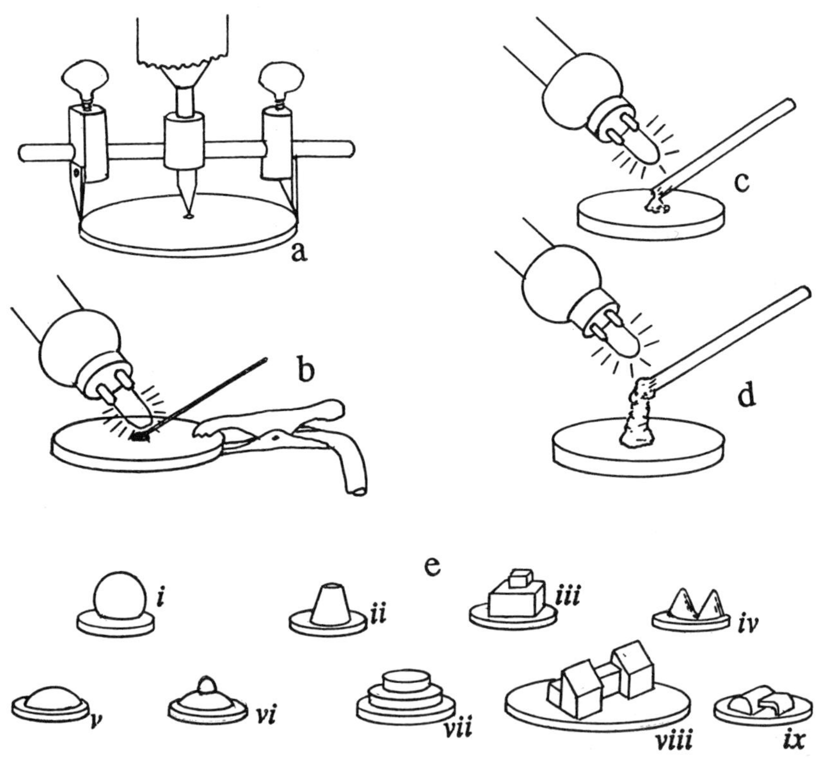

Fig. 6. Bases for miniature sculptural pieces. a. Using a gasket cutter (in drill press) to make a cardboard base, a stock of these in various sizes being kept on hand for use as required. b. Adding a drop of hot-melt glue (from a rod of extruded glue stock) to a mounting plate in preparation for forming a sculptural piece. c-d. Building up the mass of plastic from which a sculptural piece will subsequently be developed. e. Special supplementary bases upon which sculptural pieces will be formed. These bases, to be attached to a standard cardboard one, were produced by vacuum forming, blow molding, yoke/plug molding and other processes described elsewhere in the text. With the one obvious exception, the illustrated bases were formed from 0.75-inch blanks, the exception (viii) being from a 1.25-inch blank.

Precision gluing techniques

When working at the scale required in miniature sculpture, the need often arises for an adhesive suitable for bonding the various components to one another or to the base on which the assembly is mounted. Thermoplastic, or "hot-melt", glue is a superb material for the task, but to apply it with the precision required at small scale requires special preparation and equipment, glue guns—even the so-called "miniature" ones—being, of course, far too large for making the actual application. A miniature gun, mounted and used for extruding, is, however, ideal for use in producing stocks of rods, strings, and threads of glue for subsequent use on the sculptural pieces, but the electric-loop tool is then used to apply the material to the piece, the process, shown in Fig.7, analogous to that of the use of welding rods in building up metal sculpture. A wide range of stock diameters may be produced by altering the speed at which one pulls the glue from the gun nozzle once extrusion has begun. By exerting force on the emerging mass, its diameter may be reduced from that of the nozzle itself to that of a fine hair if one wishes. This "spinning" process quickly and easily produces a substantial stock of sizes which can be stored in convenient plastic packages made from polyethylene sheet stock, using a domestic bag sealer. The smaller sizes are best wrapped on small spools (turned on the lathe). Glue rods having a diameter of a millimeter or two are useful in making major bonds between the sculptural piece and its mounting plate, but for smaller components within the piece itself, strings and fine threads of glue are usually more appropriate. With the finest threads, the smallest bit of plastic visible with the magnifying lens mounted on the face of the hood can be affixed neatly and easily to the sculptural piece with the electric-loop tool and the bond subsequently worked so effectively that it can not be detected.

When one uses the glue gun to extrude the plastics used for the actual sculptural pieces, one or two supplementary nozzles are helpful (in producing rods of different diameter more easily), but these are not needed in extruding hot-melt glue, because the range of diameters extruded is easily varied by changing the temperature (by means of a variable transformer), the pressure applied to the plastic in the heating chamber (applied by means of a length of wooden dowelling from the rear of the gun), the tension (force applied in pulling the plastic from the nozzle), and the speed at which the glue is pulled from the nozzle.

To achieve maximum precision in applying the finer glue threads, one is

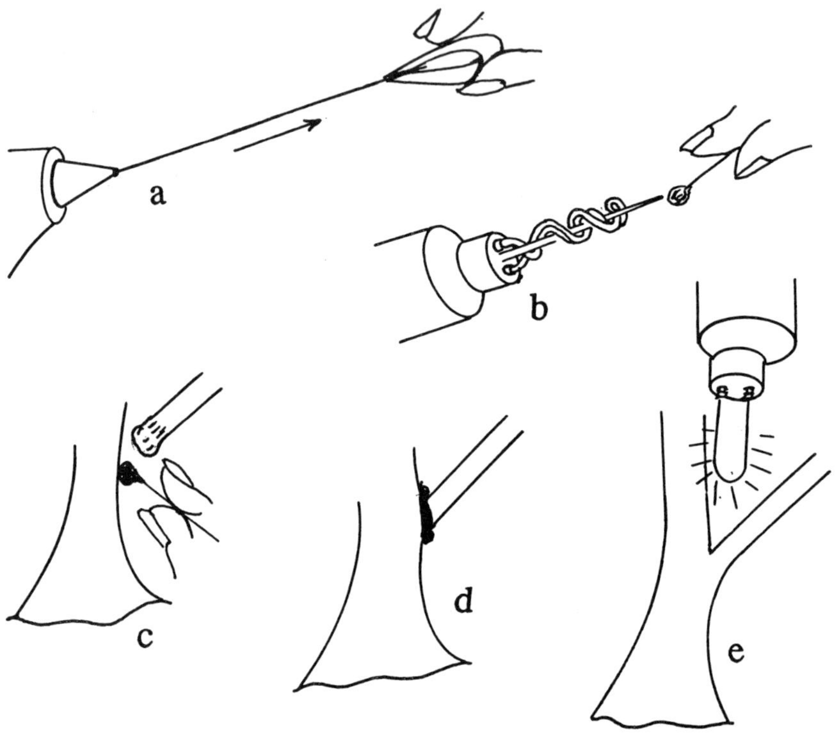

Fig. 7. Precision gluing with extruded threads of hot-melt glue. a. Producing fine threads of hot-melt glue by pulling the extrudate from a miniature glue gun (use forceps for this to avoid burning your fingers!). b. Preparing a miniature drop of glue (1 mm) by melting the tip of an extruded thread of glue. c. Applying the miniature drop to a sculptural piece; the heat-softened end of a piece of HDPE rod stock is in readiness and (d) is thrust into the heat-softened drop of glue, the whole then being fused securely to the sculptural piece with a bond so well finished as to be undetectable (e).

wise to use a supplemental tip (needle or pin) in the electric-loop tool, as shown in Fig. 7, b, but for routine use the loop itself can be used directly, the temperature reduced and the distance between loop and plastic carefully modified to suit the circumstances.

Careful control of the temperature, both of the gun and of the plastic to which the glue is applied, is generally essential in effecting a satisfactory bond without distorting the workpiece itself. This is particulary true when working with small pieces and with very thin material. A skilled worker soon learns to adjust with care the operating temperature of the loop, the distance from loop to bonding area, and the duration of heat application. Before heat is applied, one must determine whether to use the loop directly or itself wrapped around a pin or needle, the latter preferable for smaller work, of course. The addition of a pin or needle to the loop tool takes time (a minute or less), but it reduces the risk of damaging a delicate piece through overheating. With care, however, one can achieve excellent results on surprisingly small pieces and on thin film stock with the loop itself. Some of the thinnest material with which one is likely to work is that produced by such thermoforming procedures as blow-molding, yoke-plug molding, and vacuum forming, and it is with these that one must exercise the greatest care during the gluing process if distortion or melting is to be avoided. The risk is considerable, but the range of working temperatures available through the external controls of the heating loop is great, and the precision with which an appropriate temperature can be selected and maintained is quite adequate to most circumstances one is likely to encounter in these miniature sculptural efforts. As in most of the more demanding aspects of miniature work, practice is the key to the mastery of precision in gluing; once achieved, however, such mastery opens the door to a finesse essential to advanced sculptural accomplishment.

Carving and modelling with the electric-loop tool

Of the two processes basic to producing sculptural pieces, *carving* (removing material) and *modelling* (building up by the addition of material), the former can be employed on thermoplastics either in their cold or their heat-softened state. Modelling, by contrast, can only be done with the use of heat to soften the piece sufficiently to have material added and properly fused to it to form a uniform mass without tool marks. When carved cold, the material is worked with miniature tools analogous to the larger ones used by sculptors and craftspersons working with conventional materials. Small knives, chisels, files and other hand tools may be suppemented with power tools and fine burrs, cutters and sanding disks, the principal precaution in using power tools being to avoid undesirable frictional heat. (Use a speed-control device on the tool and, if necessary, a coolant.)

The process of carving heat-softened plastic with the electric-loop tool

(Fig. 8, a-g) is simply a matter of heating the loop briefly (a second or so), just enough for it easily to be used to melt and pull away a small quantity of material from the main mass and deposit it elsewhere (a waste pile awaiting reuse later, for example). The process is then repeated as often as necessary to achieve the desired form. In this way, one can do fine carving (true sculpture in the classical sense) on surprisingly small pieces. The details of the steps involved (Fig. 8, a-g) are:

1. Heat the loop (with the foot switch).
2. Melt the plastic to be removed.
3. Cool the loop (merely disengage the foot switch).
4. Use the cooling loop to lift and collect the molten plastic, letting it adhere to the loop, which is used as a rake or scoop.
5. Carry the collected plastic away, cutting the umbilical cord (which connects it to the main mass) by reheating the loop briefly and slightly.
6. Invert the loop (i.e., turn loop downward), turn on heat to melt adherent plastic, and let the plastic drip onto a cold (metal) surface, ready for reuse when, and if, required.
7. Smooth out the surface of the excision area with the heated loop.

The process of adding material (modelling) to a work piece (Fig. 8, h-l) is even simpler than that of carving; it involves these steps:

1. With the heated loop, melt a section of plastic stock (thin strip, rod, thread or string, as appropriate, volumetrically, to the task at hand).
2. Touch the molten mass to the sculptural piece at the point where the addition is to be made, using gravity and surface forces to cause the molten plastic to flow onto (and blend with) the main mass.
3. Smooth out the bonded area with the heated loop until the desired finish is obtained.

(The miniaturist interested in enhancing his bag of sculptural tools and tricks will do well to read widely and consult such books as Strong's (1938), which contain much information of use to the tool-making experimentalist. Strong's book, for example, gives the design for a simple "hot-wire" tool comparable to the "electric-loop " tool described earlier in the present work.)

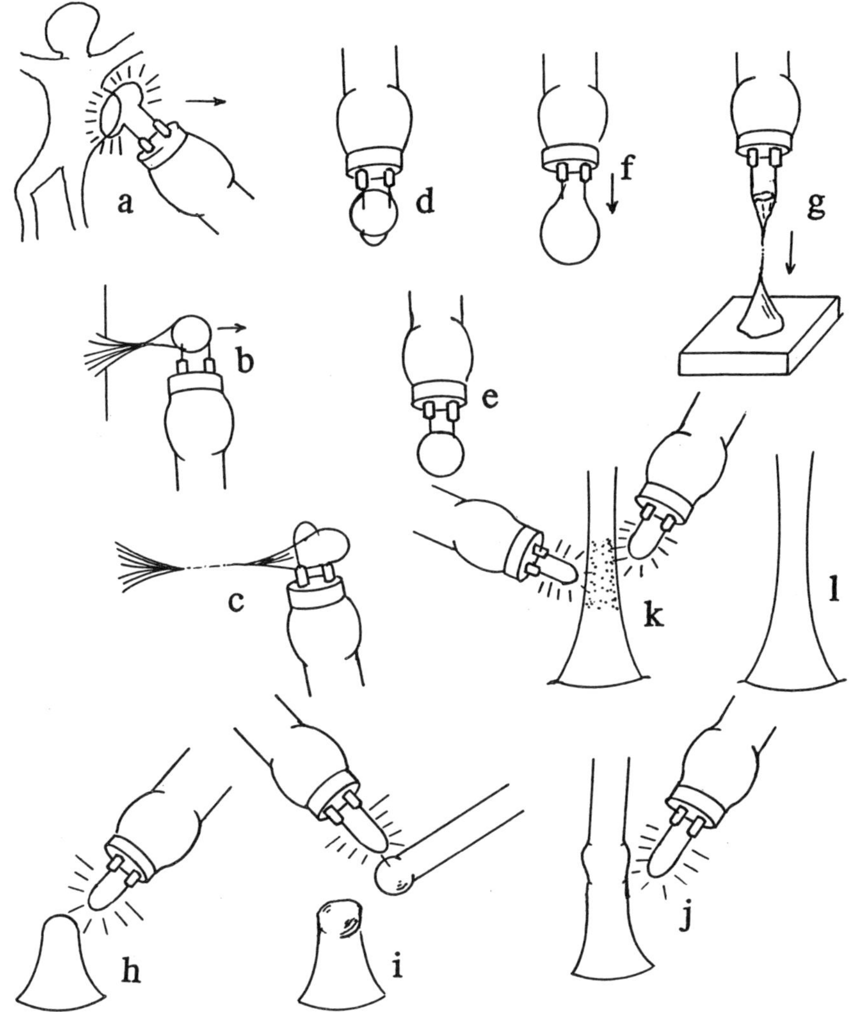

Fig. 8. Using the electric-loop tool in the basic sculptural operations of carving and modelling. a-g. Carving sculptural piece (i.e., removing material from it): a. The heated tool, its loop preshaped for the task, used to melt the material to be removed (b, c) from the main mass. d-g. The removed mass being freed from the tool by melting and letting it drop onto a cold surface (g) for reuse. h-l. Building up (modelling) a sculptural piece by adding molten plastic (h, i), joining the new stock to the older (j) and fusing the two to produce a joint so smooth as to be undetectable.

38

JOY FROM WASTE

Margarine tubs and yoghurt cups,
Have they another role to play
When once one from them dines and sups?
Why must they then be thrown away?

No, indeed, may I suggest
Another use I've found for them,
A use you'll find not second best,
But, in fact, a sort of gem?

You chop them up and melt them down,
And squirt 'em out as rods and strings,
And then these labors you can crown
By modelling lots of lovely things.

The rods and strings are blended in,
By heat applied with care,
To make wee figures fat or thin,
Or little scenes so fair.

Loose imagination's reins
And shape the plastic as you please:
Rich reward for all your pains
If you the chance but seize.

Countless sculptural gems, indeed,
In plastic waste abide,
If one but has an inner need
To express one's arty side.

What little art, indeed, it takes
To copy others' work at first,
And then improve the things one makes,
Assuage artistic thirst.

(cont. p. 39)

(cont. from p. 38)

And when at last the time is nigh
To branch out on your own,
Do not set your sights too high
As you reap the harvest sown.

Set first a goal within your reach,
Then try to stretch for one more tough.
Like this, you'll find yourself you'll teach
The skills you need quite soon enough.

Keep struggling on to learn new ways
Of developing good ideas;
You'll find the struggle will brighten your days,
Success will banish tears.

Apprehension and doubt crumble apace
As experience and skills accrue;
You'll soon delight in the developing grace
Expressed in all the work you do.

The little sculptural pieces you make
Will be a delight for all to see.
Though several hours each may take,
With each you set your spirit free,

A spirit, indeed, that roams at will
Through a miniature world of its own,
Ever free to drink its fill,
Of all the joy there known.

A harvest rich, one triply blessed:
Scrap reclaimed, all castaways,
Waste reworked, loss redressed
To brighten up our days.

PART THREE

ADVANCED TOOLS AND PROCEDURES

5

ARMATURE-MAKING

General considerations

SOME MINIATURE sculptural pieces or elements of these are more easily produced over a framework or skeleton, an armature that will support the plastic in the molten state it must characteristically assume at one or more phases in the formation of the finished work. The task, for example, of copying in miniature Maria Martins' *Rituel du Rhythm* was greatly simplified through the use of a fine copper-wire armature, it being difficult to achieve the relatively long, slender arcs without structural support during the heat-softening process required in forming them and in assembling the various elements as they were formed.

Various materials are of use in making such armatures, the most valuable being flexible wire in various relatively small gauges, twisted or not, and either single or multistrand. For some purposes, screen wire, wire mesh, and wire soldered to produce intricate skeletal shapes may be of value. Spiral scrap produced as metal waste (aluminum and brass particularly) from the lathe or drill press can be of value either directly as it comes from the machine or coated with molten plastic to match the rest of the sculptural composition (See Plate 9, b). Plant structures, such as the small stems from clusters of grapes, make good armatures for tree-like sculptural features (Plate 11, h) and can be built up in various ways with molten plastic or other materials (a dip-coating plastic, acrylic paints, or glue plus fillers, such as clay, wood flour, plaster, cornstarch). Cutouts from cardboard, fiberboard and various other types of paper or composite materials, and even such delicate materials as lace or plastic ribbon may be used if great care to avoid unintended distortion, scorching, or burning is subsequently exercised when coating them with thermoplastics. By selectively removing parts of such supporting materials with the electric-loop tool, interesting variety may be incorporated in the overall design of the sculptural piece. Ceramic clay, bisque-fired, or at least hardened well in a domestic oven or grill may be used to create a form which can then be coated with plastic, the clay subsequently broken or carved away, to leave the plastic form, the principal advantage in using clay being that it can be carved more easily than can the thermoplastic material in its cold (solid) state. Miniature topographical features, such as mountains or rolling hills may sometimes be more easily formed in clay, hardened and then coated with plastic, than directly in thermoplastic material, the coating applied either with the electric-loop tool or by blow-or-vacuum forming with the multipurpose, mouth-operated tool decribed later. Other materials, such as plaster of Paris, dental stone, and wood may at times be of value to the sculptor as frameworks for miniature sculptural pieces; a wooden armature, turned on the lathe, served, for example, as the framework on which to construct the angel shown in Plate10, g.

Of all the materials available for use in making armatures, however, fine wire (preferably copper or brass), usually two strands twisted together by mechanical means, has proved the most serviceable. This may be twisted by hand, but a special tool for accomplishing this enables one to do so more efficiently and easily, and the results obtained are generally of superior quality. Twisted wire armatures may be used effectively as supports in many sculptural pieces of a form difficult to retain in molten plastic during the formative stages and, in my experience, are generally preferred in making figures and curved forms otherwise difficult to produce in the smooth finish so often required.

A single strand of untwisted wire of appropriate gauge may at times be used instead of its twisted counterpart, the choice depending upon the circumstances and the worker's own preferences. In some cases it is necessary to preshape the armature around a suitable mandrel to achieve the exact form (particularly for curved features) required. Multi-strand wire, skillfully prepared for the purpose, can be of considerable value in supporting structures that radiate out in a single plane or several different ones from some central focal point or area. Thus, the fan-like spread of a bird's tail or the expanding form of the crown of a tree is easily produced by spreading the strands of wire to the shape desired and then coating it with molten plastic.

Wire-twisting tools

The simplest of homemade wire-twisting tools, to be clamped in a bench vise for use, can be made from a modified clothespin, a decapitated straight pin, a short length of nylon (or other soft-plastic tubing), and a small pin vise, as shown in Fig. 9, a-b. Part of one leg is removed from the clothespin so that the other can be clamped in the bench vise. The jaws of the clothespin are tapered to provide a neat holder for the length of plastic tubing (serving as a

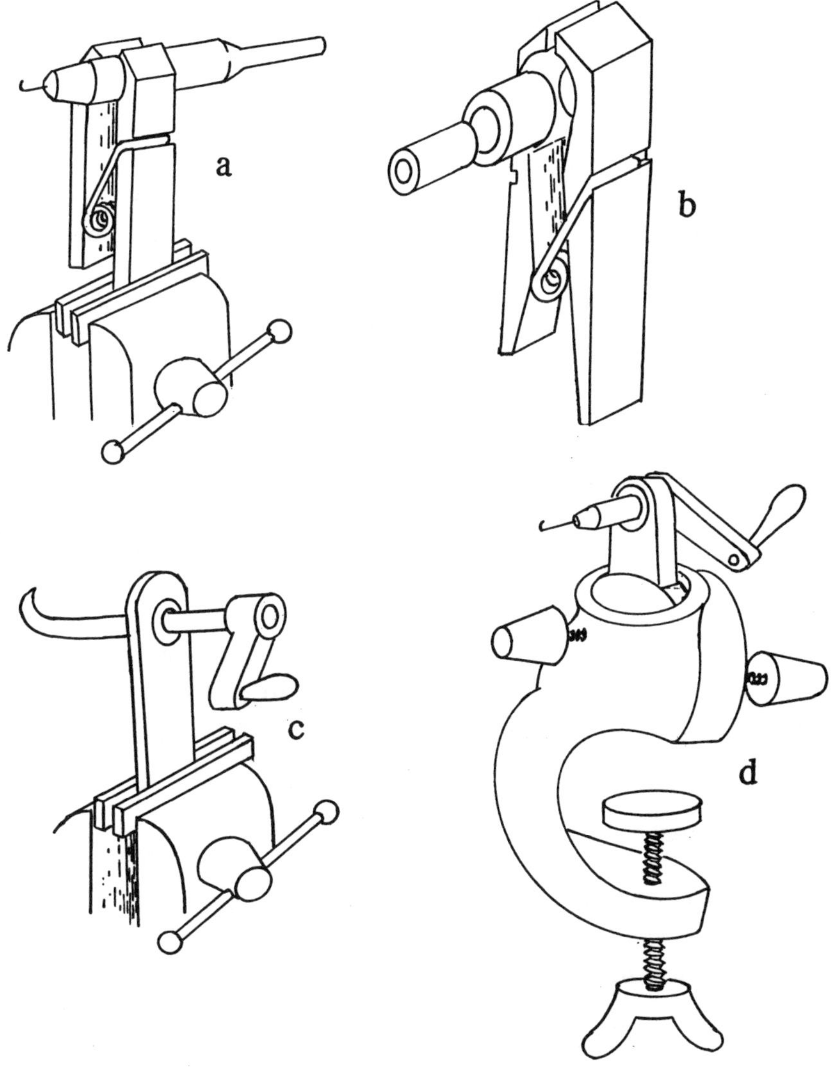

Fig. 9. Wire-twisting tools for use in making armatures of sculptural pieces. a. Simple tool made with a pin vise and clothespin. b. Exploded view to show the sleeve-bearing assembly ready for insertion between the jaws of the clothespin. c. Another simple wire-twisting tool, this one with ball-bearings. d. Tool made from scrap pieces found in a thrift shop, the ball-and-socket arrangement making it particularly adaptable and efficient in operation.

sleeve bearing) through which the pin vise (turned smooth on a lathe) passes. If the pin vise does not rotate freely in its plastic bearing, slip a small wooden wedge between the clothespin jaws to relieve the pressure slightly. A little experimentation will soon produce the desired effect. A more efficient and elegant solution, of course, is to make a metal or plastic sleeve-bearing to the exact dimensions required for efficient action with minimal friction. An alternative to the pin vise is a simple turned brass or wooden rod, tapered and drilled at one end to accept the pin (soldered or glued in place with cyanoacrylate glue) and fitted to rotate freely in a suitable metal or plastic sleeve. Useful variations on the general theme of simple wire-twisting tools, principally in the form of the bearing and its mount and in the way of attaching the device to the work bench (with and without a vise), will, undoubtedly suggest themselves to an imaginative worker.

The tool shown in Fig. 9, d, made possible by the lucky find of its basic ball-and-socket component (a lovely metal casting) in a thrift shop, is a particularly efficient and versatile one in twisting wire for armatures. The mount for the bearing, pin vise and handle was fitted into a slot (not present in the original) in the top of the ball, and a knurled screw was added (right side) to improve the clamping action of the socket. The ball-and-socket joint permits movement through a wide range, making it more convenient to use than the ones previously described; any of these take less space on the workbench than would a relatively clumsy eggbeater drill clamped in the vise.

Making armatures

The principal use for the wire-twisting tool in miniature sculpture is in forming armatures for figures of animal or human figures, although armatures may often be of use in forming other structures or features of such form that support is needed when producing them in heat-softened plastics. Generally speaking, copper wire in two different gauges suffices in making most of the armatures needed for sculptural pieces in the 1-2 inch range. At the lower end of the scale, for figures in the 0.25-0.50 inch range, a fine gauge (47 gauge) is preferable, mine salvaged from a coil of cloth-coated bell wire (the cloth removed by pulling the wire vigorously a few times through a sheath of sandpaper gripped tightly between the fingers). For larger figures, a heavier wire (23 gauge) may be more appropriate.

To make an armature for a human figure, Fig. 10, loop a piece of wire (approximately six times the length of the figure) in a V over the hook of the pin and hold both strands with a pair of pliers at a point near the waist line of the proposed figure. Produce the torso by twisting the wire a few turns in the span between pin and pliers (Fig. 10, a, *ii*). As the twisting proceeds, a loop will be formed by the pin, this corresponding to the head of the figure. Bend one of the free lengths of wire back upon itself at a point near the foot of the proposed figure. Remove the wire from the pin and place the new bend around the pin. Twist this new double strand a few turns to give the length of the leg and repeat the process with the other free length of wire to produce the second leg. The free end (single strand) of wire from each of the two legs is then bent in turn to the length of an arm, hooked over the pin and twisted as were the legs, a loop being produced by the pin during each twisting, so that the resulting armature at this stage (and after trimming away any remaining excess) resembles that shown in Fig. 10, a, *x*. These loops, or rings, become the basis for the development of hands, feet, and head on the finished figure, the wire in them bent or trimmed as necessary to produce the desired supporting structure.

Once the basic skeletal form of a structure has been produced, it generally must be built up to give the desired thickness and overall size required in the finished piece, and a surface finish must be applied to match that of the remainder of the composition. In the scale of these miniatures, the build-up is best done by melting plastic over the surface with the electric-loop tool, using rods, threads, or strings of appropriate diameter, but it should be borne in mind that this can also be accomplished in various other ways, if desired: by dip-coating in plastisol, gesso, acrylic paints, polyester resin, synthetic glues of various types, and plaster of Paris. Many of these materials, however, are easily scorched or otherwise defaced or deformed by the heat from the electric-loop tool, so it is generally best to use the same plastic as that from which the sculptural piece itself is made, i.e., either polyethylene or polypropylene.

An alternative to using a wire armature to build up a figure, is, of course, to produce the torso and limbs by welding lengths of plastic rod stock together and then shaping these with the necessary detail. Because of the size of most of the figures required in pieces of miniature scale, though, this approach is much more tedious than that using armatures. Additionally, the wire armature may easily be bent or shaped to the form required to simulate the flexibility of natural articulation required in imparting movement to a figure, a type of shaping less easily accomplished on rod-stock components

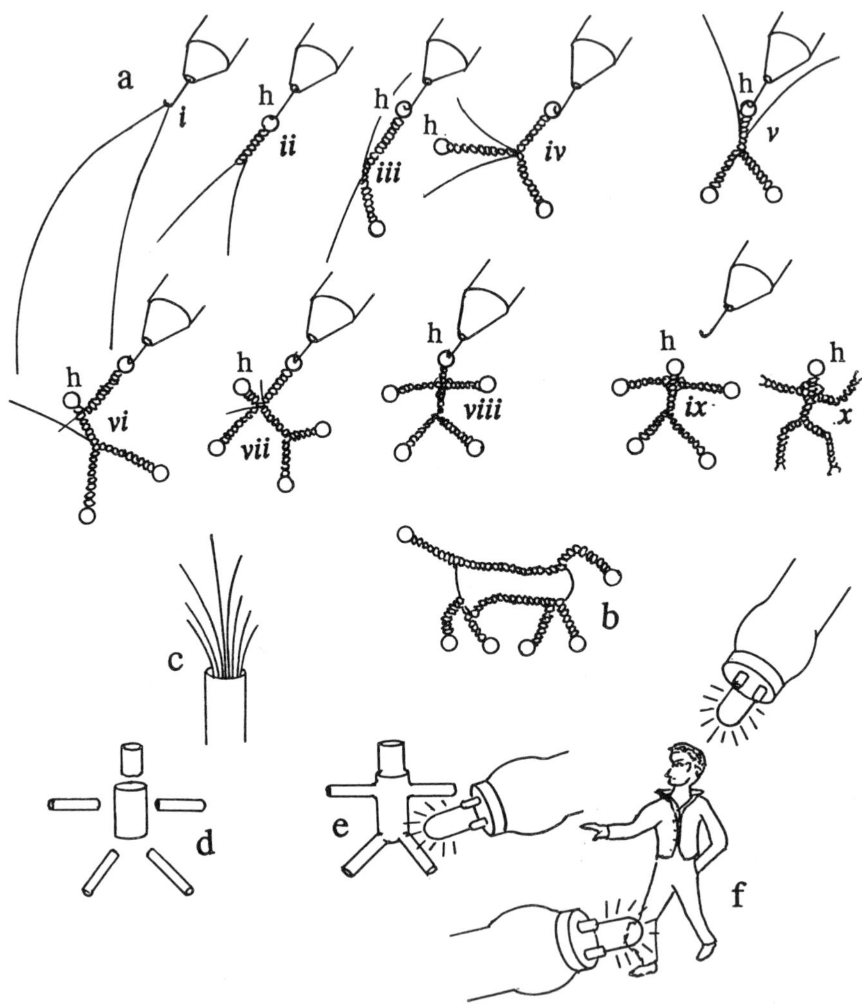

Fig. 10. Armature making. a. Steps (*i-x*) in making a twisted-wire armature for a human figure, using one of the wire-twisting tools shown in Fig. 9. See text for explanation of the various steps in the procedure. b. Twisted-wire armature for the body of a four-legged animal. c. Multistrand electrical wire is useful in making armatures for plants of various types. d-e. Alternate way of making a human figure, without the use of an armature: segments of plastic rod stock are fused together with the electric-loop tool. f. Figures produced without an armature are built up and given their final form through the use of the electric-loop tool.

lacking the support and discipline the armature provides.

Just as in stone, a medium in which it is difficult to obtain the separation of small or thin components (such as the fingers on a hand) from one another without the risk of breakage, so, too, in molten polyethylene or polypropylene is it relatively difficult to achieve the separation of such small components at a miniature scale, the reason here, though, being that the plastic tends, when heated to a molten or semi-molten state, to coalesce. By using a fine wire armature, however, this difficulty can readily be overcome. Making a suitable armature at such small scale can be a challenging endeavor, but by using a simple jig (Fig. 11) that enables one to bind the fine component wires together in what will become the palm area of the finished piece, the armature is easily made. The bases for the armature are five of the seven strands in a length of multistrand electronic wire.These, bared by removing the plastic insulation and clamped with a miniature C-clamp and a clothespin (the jaws slightly modified to improve their efficiency for this task), are soldered near the insulated end to a narrow strip of thin brass shim stock. [Miniature clamps—only 0.6 inches long—may be made from small steel U-stock trimmed to the desired width and fitted with a 3-56 thumb screw and thin spring-steel plate (Fig. 11,e). A wooden baffle strip will protect the screw from solder.] If several such armatures are being made, a small soldering iron, fitted with a special tip and modified for mounting in the drillpress chuck, may be brought down with precision onto the wires and brass strip (underneath the wires) held by the clothespin clamped in the vise on the drillpress table. In this way it is possible to produce an efficient soldered joint over which the palm of the finished hand will be modelled, all at a scale commensurate with that of the entire figure one will form. [A simple way of making small uniform pieces of solder for such joints is to crimp a length of solder wire tightly between the parallel-scored jaws of a pair of pliers and sever the resulting embossed segments with a knife.] When the copper strip has been soldered to the wires, they can be cut free from the insulated section and trimmed to finger length. Because of their diminutive size (hand length: 0.25 inch or less), it may still be so difficult to coat them with molten plastic without coalescence that resort to acrylic paint, applied with a fine sable-hair brush, is the logical alternative.

Another way of producing minute hands, this time without the use of an armature, is to roll a knurled tool over a heat-softened strip of thermoplastic in the way shown in Fig. 31, b. The tool, merely a toothpaste-tube cap fitted with a wooden handle (glued in place with epoxy cement), is pressed firmly

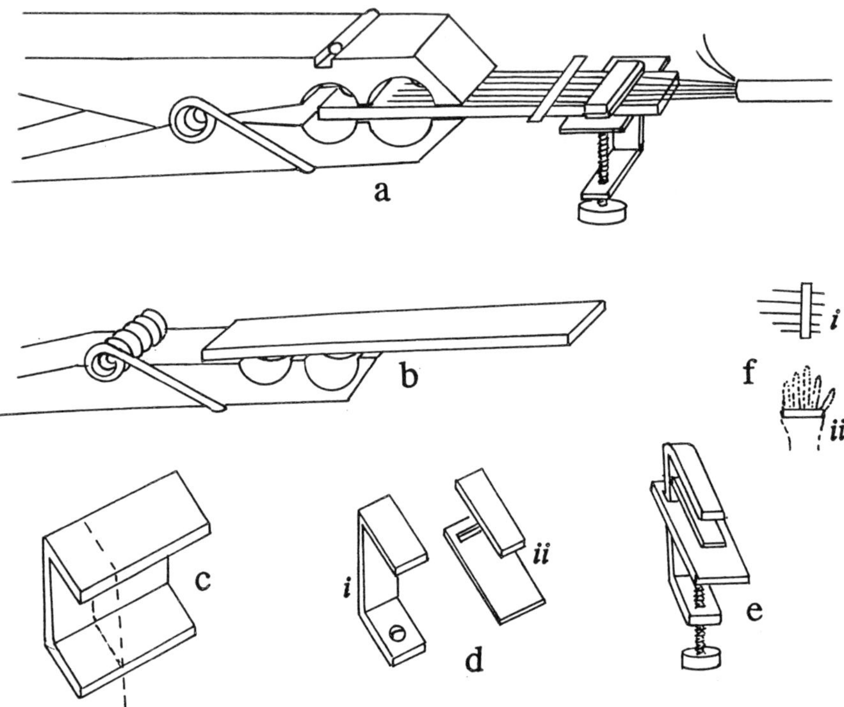

Fig. 11. Jig for making armatures for hands and feet. a. The jig in use, ready
for soldering the elements to produce the basic armature (f, *i*). The component
parts of the jig are shown in b-e. b. The lower jaw of the clothespin with wooden
platform strip attached (glued). c-e. The body of the miniature clamp (0.6 inches
tall) is cut from steel channel stock and tapped (d, *i*) to receive a thumb screw. d, *i*.
A moveable metal plate (cut from scrap metal strap) and a hardwood strip (d, *ii*) fit
as shown in e to complete the clamp. f,*i* The finished armature for a miniature
hand (0.25" long), showing copper (or brass) crossbar which holds the finger
components in place and serves as a base for developing the palm and wrist of the
finished hand (*ii*), built up with plastic, using the electric-loop tool.

into the soft plastic while being rolled along the strip, a marginal portion of the latter left intact (Fig. 31, e) to serve as the palm when individual hand units are finally severed from the strip. It takes merely a few finishing strokes with the electric-loop tool to complete the final shaping of fingers and palm after the unit has been attached to the arm of a figure. (The hand-making procedure is treated further on pp. 104-5.)

Fan-like structures, including the fantails of birds (see Plate 7, e, k) may be made from multistrand electronic wire merely by stripping the insulation from a length córresponding to that of the tail, fanning the strands out, and severing the entire unit from the insulated stock, leaving an insulated section of about two millimeters to serve as a base for attachment to the body of the bird. Care should be taken to leave enough insulation intact to prevent separation of the wires after the unit is cut from the stock piece (a drop of cyanoacrylate glue here is good insurance).

ARMATURES

Plastics, when softened or melted by fire,
Are quite well supported by twisted wire
That gives them strength and helps them endure,
Merely a twisted-wire armature.

The wire is obscured by the coating of plastic—
A simple task and not at all drastic.
As the wire disappears, hidden from view,
Its job now done, we bid it "adieu",

But we know it's still there, playing its role:
Helping the sculptor maintain control,
Giving its strength, though buried within
The masses of heaped-up plastic-skin.

So, here's to the armatures of wire;
We know they will do the job we require.
From us all they ask from their hidden retreat
Is a job well done, tidy and neat.

6

MAKING URSCHLEIM AND ANLAGEN

THE THREE BASIC WAYS of producing miniature sculptural pieces in thermoplastics are:

1. Carve from a solid mass (the method of subtraction), using the electric-loop tool.
2. Build up by the accretion of components (the method of addition), using the electric-loop tool.
3. The relatively aleatory method of developing from a nondescript primeval molten mass *(Urschleim)*—by pulling and distorting it in any one of a variety of ways—a single or a group of suggestive forms *(Anlage, Anlagen*, respectively) from which a final form can, by the conventional methods of sculptural addition and subtraction (no. 1 and 2 above), be produced.

Generally speaking, in the first method one starts with a solid mass or with solid components and merely removes unwanted material to produce the final form. In the second, starting essentially with nothing but isolated bits of stock material, one builds up a piece by adding bits of stock material to one another. The third method, by contrast, starts with a *molten* mass of plastic which can then be worked in various ways to produce the prefigurement of the finished piece. This prefigurement, or Anlage, can then be further modified in either of the first two ways to produce the final piece. Borrowing German terms and concepts from geology and biology, the initial molten mass [comparable to the *Urschleim* from which 19th-century science considered all life to have arisen; the term here anglified (no initial capital)] is deformed—or, more properly, *preformed*—through various techniques to produce a wide variety of shapes having the potential (and thus analogous to the *Anlagen* of the embryologist) of further development into finished pieces at the hands of the sculptor.

Strictly speaking, and in analogy with embryological usage, the anglicized term *anlage* (plural= *anlagen)* can be applied to any undeveloped mass that will be the foundation or rudiment of a sculptural piece. I propose, however, to apply it more restrictively, namely, to a precursory or embryonic form produced from a molten mass by any one of several different processes used either singly or in conjunction with one another. By analogy, then, the extruded stocks of plastic from which finished sculptural works can be

50

produced through the usual additive and subtractive sculptural procedures are not, properly speaking, anlagen, just as living cells in general, though the building blocks of tissues, structures and organisms, are not termed "anlagen" except when they are encountered in the developmental stage (i.e., in cell multiplication) following the fertilization of an egg by sperm.

Remembering from my youth the delightfully intricate and delicate forms we lads produced by pouring molten lead into a bucket of water, and hoping that molten polyethylene would do the same, it came as a crushing disappointment to see that this material, when handled in the same brusque manner, responded in a most coarse and uninteresting way, possibly because its inferior specific gravity prevented it from falling through the water column and, in effect, exploding as a lovely inflorescence as lead does, but, rather, merely floating on or near the surface as a most uninspired and uninspiring formless blob. For the same reason, this material, when allowed to fall free in air and land on a metal plate, tends to produce mere rounded blobs, shapes of little use as anlagen suggestive of imaginative sculptural development. The initial disappointment with these early attempts at producing potentially useful anlagen in such simple ways led to experimentation along different lines, the description of the more promising of these being the subject of the remainder of this chapter.

The sequence of steps involved in producing a sculptural piece from an undifferentiated mass of molten plastic is shown in Fig. 12, a-d. The initial molten mass is produced by melting scrap plastic of suitable volume over a gas burner (laboratory burner of the Bunsen or Fischer type, or a propane torch—*extreme care being taken at all times to avoid igniting the plastic and to have proper ventilation in the work area; keep a fire extinguisher handy and take all necessary precautions to avoid fire and to have a safe exit at hand should an accident occur. This is a potentially hazardous activity!*). I find a microbiological burner, the type with a small pilot flame and a wrist-contolled valve, particularly convenient. An electric fan, placed nearby on the work bench, but so positioned that it does not affect the flame itself—only the air current over the work area—carries the fumes away from the operator.

To make the initial mass of molten plastic, a strip of plastic cut from a yoghurt container or other handy source is gripped at one end by a pair of pliers (or a special work positioner, as shown in Fig. 12, e) and the opposite end is carefully heated until enough of it is melted (avoid igniting the plastic in the process) to provide the desired volume. This molten mass is then

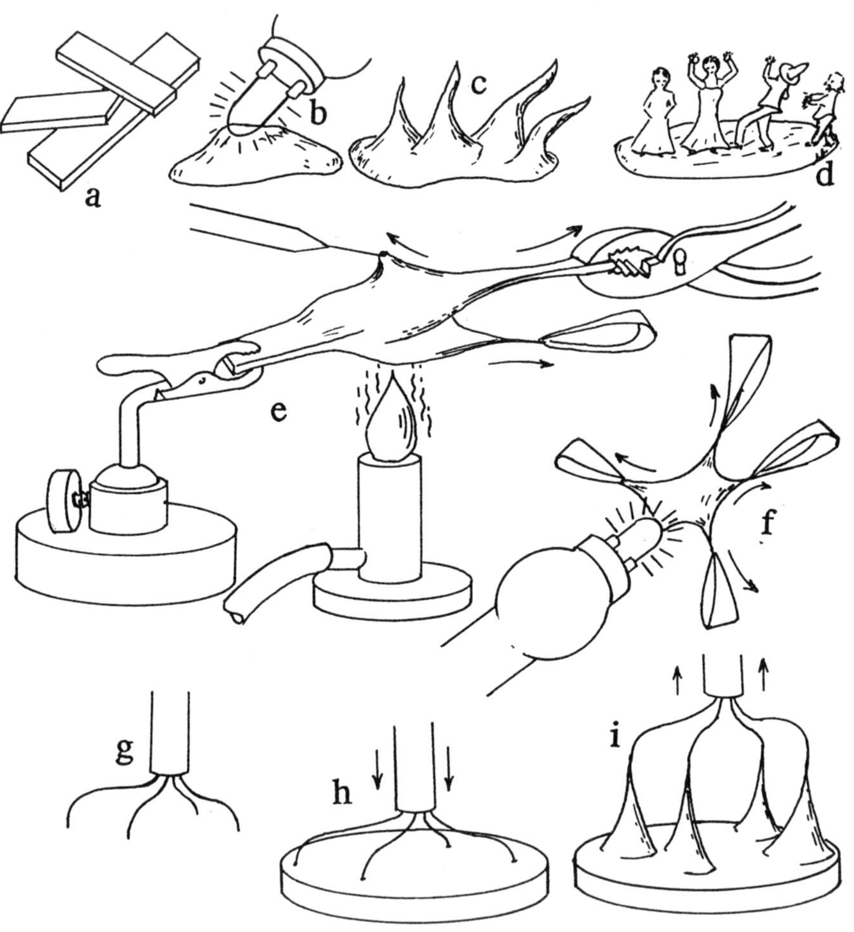

Fig. 12. Producing the urschleim and anlagen from which a miniature sculptural piece will be developed. a. Stock strips of plastic (HDPE, cut from yoghurt containers). b. Melting plastic to produce the mass (urschleim= primaeval ooze) from which the anlagen will be formed. c. The anlagen (sing.: anlage) from which a finished sculptural piece will be created. d. Miniature sculptural piece (0.75" high) developed from such anlagen as those shown in *c*. e-i. Procedures used in producing anlagen from a molten mass (urschleim) of plastic. e. Pulling out a heat-softened mass of plastic with pliers, tweezers, and a needle tool. The mass to be heated is relatively large, so a miniature Bunsen burner is used

quickly transferred to a piece of metal clamped in the bench vise, the piece unheated and large enough to absorb quickly the heat from the plastic and dissipate it sufficiently rapidly for the under surface of the plastic to adhere to it as the upper bulk of material is pulled or otherwise deformed to produce the anlage. By placing the heated plastic on a cool, heat-absorbing metal surface, a viscosity gradient, ranging from rigid solidity to near fluidity, is quickly established, this simultaneously establishing a corresponding gradient of resistance to deformation by external stress or tension. The operator takes advantage of these two parallel gradients and uses them in deforming the urschleim to produce the wide range of shapes achievable in forming the anlage(n). The extent or range of the viscosity gradient is directly proportional to the difference in temperature between the molten mass of plastic and the metal plate onto which it is placed, and, since the range of forms achievable by deformation through externally applied force is directly proportional to the range of this viscosity gradient, it follows that the temperature contrast should be as great as possible (i.e., without resort to the use of refrigeration and dangerously intense heating). Were the metal plate itself heated at the outset, the working time for the operator before the plastic solidied would be increased, but, most importantly, the viscosity gradient would be restricted commensurately and the gradient of resistance to deformation would be reduced, this reducing the range of forms one is able to produce from the mass.

The initial deformation of the urschleim of molten plastic to produce an anlage can be achieved in either of two ways, i.e., passively, through gravitational flow, and actively, through counter-gravitational pulling or drawing. Actually, each of these is employed in most anlage-producing operations, for, to achieve the best results (i.e., the most imaginative forms), some anti-gravitational force is usually combined with the passive draping action that results from the simple gravitational flow of the molten mass, and, conversely, a great deal of gravitational flow can be permitted (and is, in fact, highly desirable) in the case of counter-gravitational deformation.

The entire process, from urschleim to finished piece, can be executed

---

for the task. (*Observe proper safety precautions when heating.*) f. Melting and pulling a smaller mass of plastic. The electric-loop tool is an adequate heat source for this, the miniature tweezers used to pull sections of heat-softened material in succession. g. The splayed strands of wire, exposed by removing some of the insulation from multistrand electric wire, make a useful tool in pulling heat-softened plastic (h) into anlagen (i) for use in creating miniature sculptural forms.

through the use of the basic or minimal facilities described in Part Two, but unless an inordinate amount of time is to be spent in producing larger and more complex pieces, it is wise simply to expand the facilities sufficiently to permit one to melt somewhat larger masses of plastic and then deform them quickly and in more varied ways. This essentially means merely the addition to one's facilitites of a gas burner, a sturdy work-holding device (bench vise, for example), and a pair of pliers.

The aim of one's work on a molten mass can be either to produce the anlage of a single figure or structure or several anlagen from which several figures of more complex structures or forms can be developed. With the simple facilities just mentioned, one can produce a useful array of relatively simple anlagen, the pliers or forceps used to pull part of the plastic from the molten mass. If, however, one wishes to produce more complex anlagen, it becomes helpful to have some simple mechanical aids to the multiple drawing of material from the molten mass, the principal difficulty to be overcome being that the plastic cools so rapidly that any deformation must be quickly done. Using tweezers alone, and working with both hands simultaneously (the work piece held by a work positioner, as in Fig. 12, e), it is difficult to pull more than two different areas of the molten mass before the mass cools.

The tools shown in Figs. 13 and 14, however, enable one to pull the plastic from various points in rapid succession. The first of these consists of four pull rods mounted in simple friction sleeves atop flexible arms attached to a metal baseplate. The arms are simply lengths of insulated 12-gauge electrical wire, one end inserted into a hole drilled in the baseplate, the other stripped, flattened, and soldered to the brass sleeve-bearing in which the pull-rods slide. If one wishes greater freedom in the anchorage of the flexible arms, a magnet can be used as the base of each and this placed on the baseplate wherever fancy dictates. The brass pull-rods are fitted with a turned knob at the free end, while the working tip is either hooked, flattened, roughened (to facilitate adhesion), or left unmodified but capped with a tightly fitting length of small-diameter nylon tubing (this to facilitate freeing the rod from the plastic after pulling is complete). For use, the flexible arms are so positioned that each of the pull-rods in turn can be inserted into the molten plastic and then pulled upward to bring with it a quantity of the material the desired distance and then left to cool. If, after cooling, the plastic cannot simply be pulled free of the rod, the electric loop tool can be used to heat the rod, facilitating removal. If a rod with a nylon sheath has been used, the rod can be pulled free of the sheath, leaving the latter attached

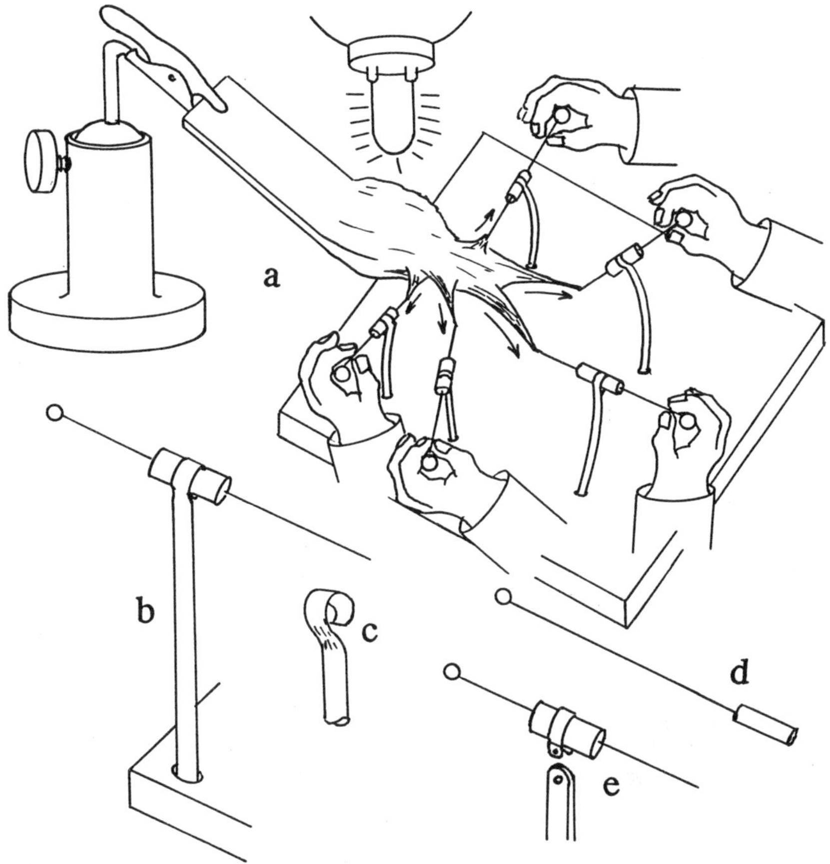

Fig. 13. Set-up for pulling anlagen. a. Plastic stock, held in a work positioner, is heated with the electric-loop tool, and heated sections are pulled in succession, using special pulling rods (b) mounted by flexible shafts to a base board. b. Pulling tool and mount. The pulling tool consists of a stiff wire passing through a sleeve bearing and attached to the flattened end (c) of a length of heavy-gauge electric wire, the lower end of which is press-fitted into a hole drilled in the baseplate. By slipping a short length of nylon tubing over the free end of the pull-rod (d), the molten plastic, when cooled, can more easily be separated from the pull-rod than if the bare rod were adhering to it. e. Suggested construction for a bearing mount that pivots (thumb screw for tightening it in position is not shown).

to the plastic until such time as it can be freed (also with the electric-loop tool if necessary); by using sheathed tips, one batch of shaped plastic can be set aside and the way cleared for making another, the pull-rods simply ensheathed in fresh nylon tips from stocks prepared in advance.

By mounting both the work piece and the heating tool (in this case the electric-loop tool) in holders that free both hands for the pulling operation, and by sustaining the application of heat, the plastic mass can be kept molten for protracted periods, and more complex pulling and shaping can be achieved than when the mass is allowed to cool directly after having been melted. An advantage of sustained heating, particularly when the heat can be concentrated on local areas within the pulled mass, is that one pulled section can be heat softened and pulled again.

While one can, with the sliding-rod tool, produce fairly elaborate anlagen, these are somewhat stereotyped by the fact that the rods slide in fixed bearings and can easily move only in straight lines, the consequence being that the plastic is pulled in straight directions. Although the position of the bearing (and, thus, the direction of the deformation achieved with the sliding rods) can be altered (the support arms being flexible), it is not easy to change this position during the deformation process, because one must work rapidly and is generally too busy pulling the several rods (before the plastic cools) to be able to shift the position of the tool arms as well.

Further diversity in the pulled products can be achieved through the use of dop sticks operated freehand (Fig. 14, b-e), the tip of each inserted in turn into the molten mass, pulled out to produce the desired form of plastic, and the opposite, or free, end of the stick then affixed (by a magnet or blob of wax attached to it) to the tool's baseplate. If instead of a simple flat baseplate one uses an L-shaped one (Fig. 14, g), the form of anlagen produced will be varied, those produced with dop sticks attached to the horizontal base section different from those produced with dop sticks attached to the vertical section.

Another way of producing multiple or complex anlagen is to press the tips of multistrand wire into a molten mass of plastic and then pull the parts of the attached plastic upward with the wire, the principal limitation of this method being that the anlagen so produced tend to be similar in form (though some may be larger than others) as a consequence of the fact that the lines of force act essentially parallel to one another in spite of the radiating form of the wire-strand mass. Instead of multistrand wire as a pulling tool, one can use (Fig. 14, h) nails, toothpicks, or short lengths of bamboo skewers attached to a wooden plate (serving as a handle). Though some variety may be achieved by twisting the tool or pulling it in an arc before the adherent plastic cools, the individual anlagen tend to remain similar to one another in form, though not necessarily in size. After cooling, the plastic can be removed from the elements of the tool by heating with the electric-loop tool.

A particularly useful procedure for forming anlangen is one here termed the *pulled-adhesion* process. This involves merely the sandwiching of a molten blob between two blocks of wood [or sturdy cardboard (mounting board, railroad board, etc.)] fitted with short handles (dowelling), pressing the two blocks together (to assure adhesion of the plastic to them), and then pulling them apart the distance necessary to produce the desired dimensions. The plastic is drawn out between the two blocks and forms an interesting mass of drawn-out connections, in form roughly analogous to the stalactites and stalagmites of cave deposits, imparting to the overall structure the appearance of the interior of a cave (see Fig.15, g and Plate 7, b). The resemblance is, of course, purely superficial, because the structures in the two situations are produced by processes that are radically different, those of the plastic by tensional forces acting to dismember a molten mass, those of the cave by gravitational forces acting on the dripping of percolating mineral-laden groundwaters. A principal difference between the end products of the two is that whereas the size of the stalagmite (the lower deposit in a cave) increases with time and tends to be come larger or more massive than the stalactite, the lower and upper components of the pulled plastic tend to be similar in both size and form.

As shown in Fig. 15, a, the end portion of a strip of scrap plastic is heated over a Bunsen burner or propane torch to produce the initial molten blob. Small rectangles of Philippine mahogany are particularly favorable in making efficient tools for the adhesion-pulling process, because the open grain of this wood assures good adhesion, yet the wood can easily be separated from the finished plastic piece with a thin flat-bladed tool (a putty knife is good for this); the handle is attached (glued and nailed) to each of the rectangles. It is wise to have several such tools on hand so that they can be used in succession, each pair being set aside to cool while others are in use.

The plastic selected for the procedure should be one that will, when molten, adhere well to the wooden tool and have good drawing properties, i.e., it can be pulled out into long flexible strands or threads with a pin or needle. These properties may be intensified by adding some hot-melt glue to the molten mass or melting some along with the mass so that they are blended

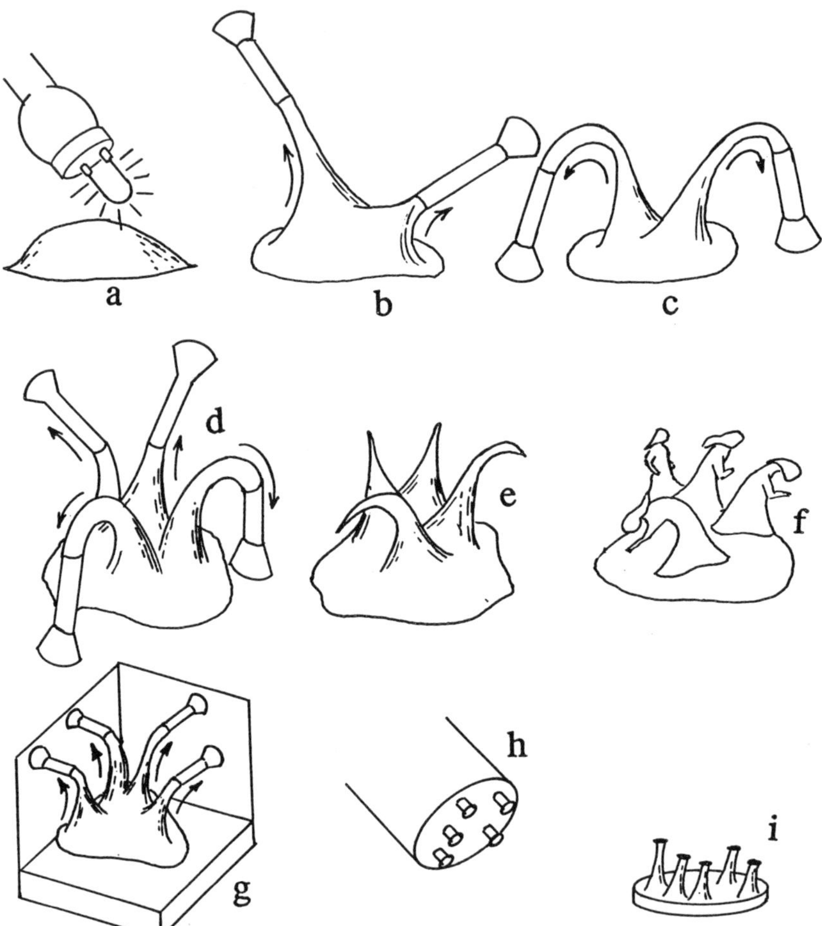

Fig. 14. Additional tools and procedures for making anlagen for miniature sculptural pieces. a-g. The use of dop sticks to draw out anlagen. Short lengths of wooden dowelling are pressed (b) into a molten plastic mass (a) and then drawn upward and then downward until the free end (fitted with a blob of wax) can be pressed securely to the work bench or baseplate until the plastic solidifies in the shape thus imparted to it (c). Two or more dop sticks can be so used in succession (c-d) to produce the desired number of anlagen (e) from which the sculptural piece (f) will be developed. By using a workboard having walls (g), the dop sticks can be attached to the walls after pulling; this produces anlagal forms of greater diversity. h. A pulling tool consisting of a length of dowelling with several small nails driven into it, used to draw several anlagen simultaneously from a molten plastic mass. After pulling, however, the plastic around each nail must be melted (with the electric-loop tool) to free the mass (i) from the tool.

well during the sandwiching process.

[Precautionary note: When melting the plastic, one should work in a well-ventilated area, preferably with an electric fan or blower to remove the fumes, and with a fire extinguisher close at hand. Considerable care must be exercised at all times to avoid igniting the plastic and also to avoid having molten plastic contact the skin, burns from it being unpleasant. This is a potentially hazardous procedure and should be treated with the caution and respect it deserves!]

This pulled-adhesion procedure lends itself to the creation, not only of cavern-like scenes, but also to the production of miniature stage-like settings. [Miniature crêches, like the one shown in Plate 10, f, measuring less than an inch in height and peopled with wire-armature figures less than half an inch tall are a popular item to pass out to one's friends.] By using arcuate, rather than rectangular, sandwich boards, a stage-like base and roof for the finished plastic piece is easily produced, but in any case, of course, it is a simple matter to modify the piece with the electric-loop tool when the rough anlage is being worked up into a finished sculptural piece. Often, too, one is able to excise one or more components of a drawn piece and use them as anlagen in the development of yet another composition. A stock of such pulled-adhesion pieces can be a treasure trove for further creative effort.

Interesting diversity can be achieved by modifying the pulled-adhesion tooling and procedure as follows:

1. Prepare a rectangular baseplate having a nail at each corner for another (upper) plate to rest upon.
2. Sandwich a molten blob of plastic between the two plates.
3. Press the plates firmly together (to assure good adhesion) and then pull them apart, rotating them 90 degrees in the process so that the uppermost one comes to rest atop the nails at the corners of the lower one.
4. While the plastic is still soft, deform its various elements with a needle or stiff piece of wire. (The plastic cools quite rapidly, so one must be modest in one's effort but work quickly, not attempting more than can be accomplished in the brief interval allowed by the cooling process.)

Another class of procedures for forming anlangen—this one combining drape forming with pulling—is based on the simple gravity flow of molten plastic. Interesting diversity in form can be achieved, for example, by letting

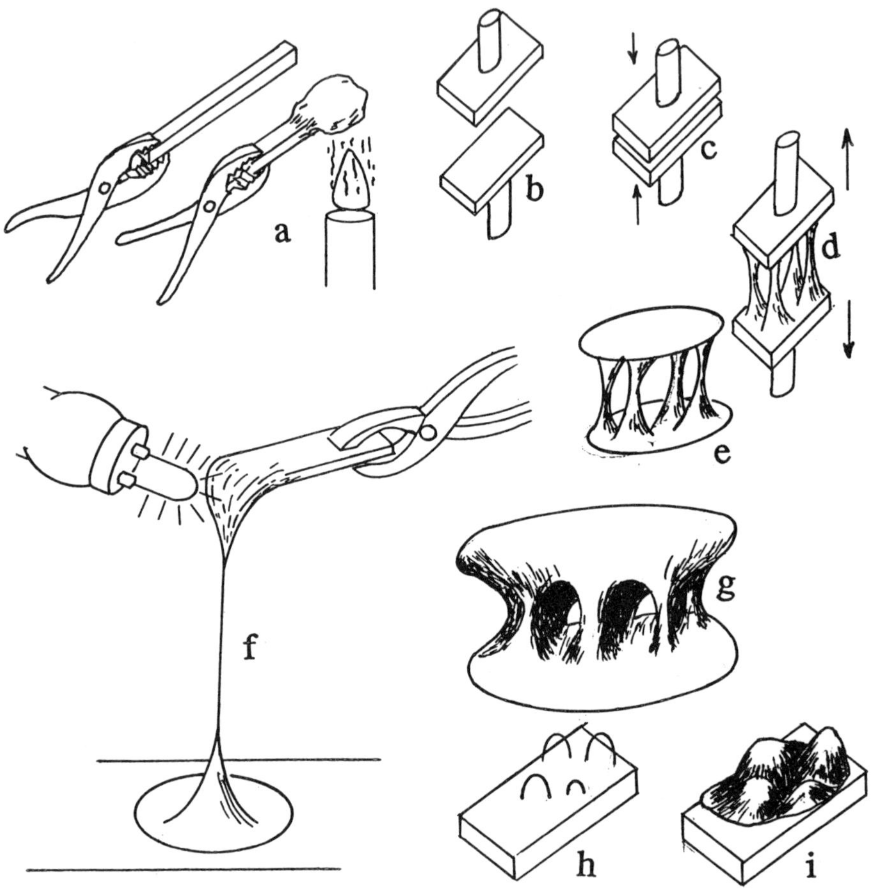

Fig. 15. Urschleim and anlagen production by the pulled-adhesion process.
a. Stock plastic strip being melted at one end in preparation for sandwiching
between two wooden plates (b, c) and then drawn apart (d) to produce four anlagen
(e) from which a miniature sculptural piece will be developed. f. A simple way of
producing a relatively small molten mass (urschleim) is to heat one end of a strip
of plastic and let the molten part flow down onto a cold surface, at which point an
anlage can be formed for further heating and pulling (with tweezers or a needle
tool). g. Cavernous anlage of the type easily developed through the pulled-adhesion process shown in a-e. h. A small wooden plate with bent nails driven into it
serves as a framework on which to develop an anlage for a bit of hilly terrain (i) by
draping a heat-softened sheet of plastic over it.

the plastic mass flow onto a simple framework of bent nails (constructed atop a wooden block) and, as it flows, pulling parts of the mass outward or upward with forceps, a needle or a length of stiff wire. A variation in the procedure is to let the plastic flow over an armature developed from a length of multistrand wire. Spraying the wire with a mold-release agent in advance will simplify its subsequent removal from the cooled plastic, on the assumption that the plastic is allowed to coat only the extremities of the armature, but, if one prefers, the wire can be left in place, the exposed parts simply nipped away from below.

Solid patterns, carved or milled from hardwood (boxwood is excellent for the purpose), metal, plastic, plaster, wax, or clay and then reproduced in plaster or polyester cast from plaster, latex or silicone-rubber molds, can be draped with molten plastic to produce the anlage of a desired form that cannot be produced in one of the simpler ways; this more elaborate approach can be particularly valuable if replicated or modular forms must be made.

URSCHLEIM AND ANLAGEN

Urschleim and *anlagen* and such words as that
Help to add interest to a subject that's flat.
Teutonic monsters to you though they seem,
Real sparkle they add to our discursive scheme.

Just think where we'd be if we hadn't these two,
The old words we'd use instead of the new:
A primaeval ooze or *plastic blob,*
Both for the first are inept at the job.

Rudiment or *foundation* or *first massing of stuff*
For the second, I fear are still not enough,
For circumlocution and ambiguity
Are poor substitutes, just senseless vacuity.

So let us adopt each Teutonic waif;
Properly defined their use shouldn't chafe.
Perhaps we can add them to the jargon of sculpture
Without real risk of linguistic rupture.

7

THE USE OF MANDRELS

THE PROBLEM OF forming graceful curves, coils, or bends in rods, strings, or threads of plastic stock or in other elements to be included in a sculptural piece through the use of the electric-loop tool without deforming the part itself or other nearby parts of the piece in the process can be a difficult one. In some cases, the use of an armature within the piece solves the problem, but in others an alternative forming procedure is desirable. The obvious approach of preforming the element separately and then affixing it to the developing piece does not solve the problem, because the electric-loop tool is not of itself suitable for producing these more demanding forms. As soon as a small element is heated sufficiently with the loop to cause it to bend, gravity pulls the softened material downward in a sharp angle rather than in the gentle curve one would like, the need quite obvious for a mandrel or form around which the desired shape can be formed, in which case either the plastic itself or the mandrel could be heated. The simple setup shown in Fig. 16, a will enable one to employ either approach, sufficient heat from the electric-loop applied either directly to the plastic piece to cause it to collapse over the mandrel (around which it can then be wrapped as required) or to the mandrel itself, in the latter case the plastic heated by the mandrel and wrapped until it assumes the required form.

Relatively large pieces, i.e., ones that can be manipulated easily and hand-held over a small laboratory burner, can be shaped either freehand or over a cold mandrel, then cooled rapidly (dowsing in water) to prevent the backlash of "plastic memory", but for pieces too small for such relatively coarse manipulatory procedures a heated mandrel can be a boon. (A major disadvantage of freehand shaping is that the plastic expands and contracts in ways that are difficult to control, hence the usual resort in commercial operations to the use of molds. The various applications of molding procedures and simple molding tools will be treated in Chapter 10.) Some fine string or thread stock can effectively be formed over a cold mandrel without even heating the stock, particularly if relatively large curves are the desired end, but when tight ones, such as spring coils, are to be formed, a heated mandrel is essential.

62

Whenever heated mandrels are to be used, it should be understood that they perform best if the stock being worked is relatively thin and small, because for best results the distribution of heat through the piece must be so uniform that the entire piece is made sufficiently soft for shaping before any part of it (particularly that directly in contact with the mandrel) actually melts. A critical point here, of course, is so to control the mandrel temperature that the softening point, but not the melting point, is reached. This becomes more difficult as the cross-sectional dimension of the piece increases. By operating the electric tool through a variable transformer, its temperature can be controlled with precision and so set that the critical temperature is reached and held. A simple way of roughly indicating the mandrel temperature itself (as contrasted with that of the heating loop) is to incorporate a depression for a few drops of water (or merely wrap a string around the mandrel and soak this with water). When the water boils, the danger point is nigh. A small quantity of the plastic stock itself, when placed on the mandrel, can serve to indicate the melting point, as can a drop of wax (wax crayon) or the commercial product Tempilstix.

A more efficient approach when working with relatively small elements, however, is to use a mandrel attached to and heated integrally with the electric loop. As mentioned earlier, the electric-loop tool is designed to accept a straight pin (decapitated) around which the loop can be coiled to heat it. By the simple expedient of drilling other holes in the head of the tool, one can attach mandrels of different diameter to it and use these in forming the plastic elements, many of which fall within a size range appropriate to the mandrels that can be fitted into the restricted space of the tool head. If one elects to rely on this approach alone, it would be wise to designate one electric-loop tool specifically for this type of operation and have it readily available within the hood. In order to free both hands for manipulating the plastic material, the tool should be clamped securely in a stable holder at proper working height (Fig. 16, b). Additionally a positioner and holder for the work piece (small rod stock, string, or thread, e.g.) should be provided so that one end can be held securely while the rest is being shaped on the mandrel. With this arrangement, it is an easy matter to form tightly wound coils in the finest of plastic threads, the temperature of the mandrel carefully reduced appropriately by means of the variable transformer in the control unit. As soon as the coil is formed, the heat is turned off (foot switch) and the finished piece—quickly cooling—can be removed and added to the sculptural composition of which it will become a component part.

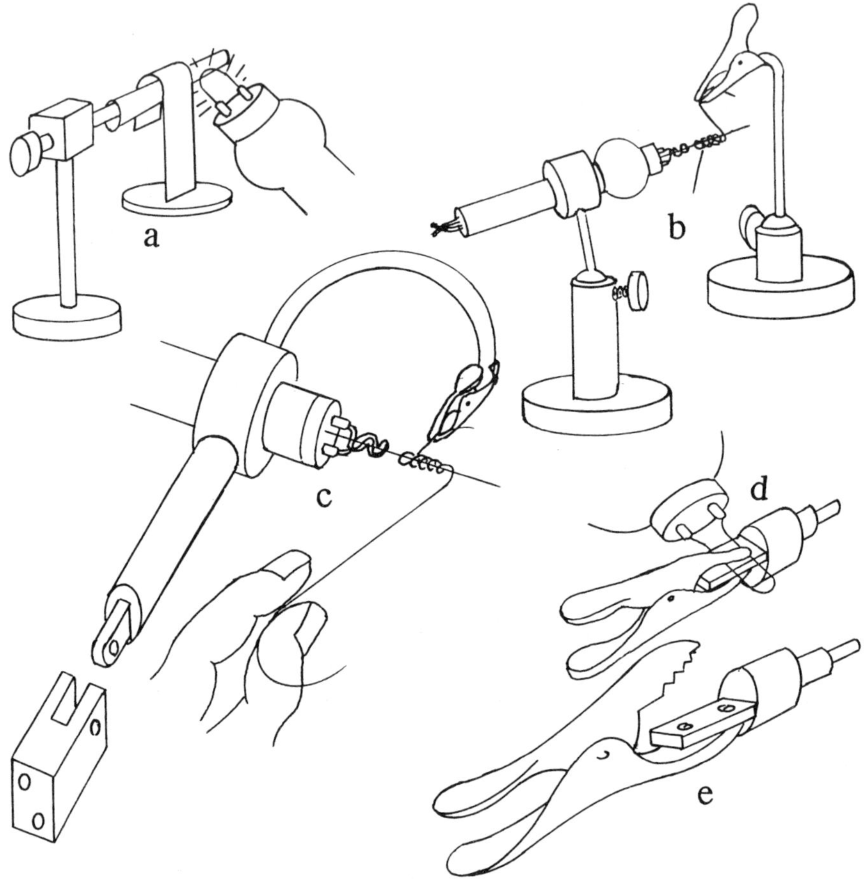

Fig. 16. The use of mandrels in shaping plastic components of miniature sculptural pieces. a. Heating a small tubular brass mandrel over which a strip of plastic is being shaped, the mandrel supported on a movable positioner, the plastic affixed to a simple acrylic or heavy card disk. b. Extruded plastic string being twisted in a coil over a mandrel (pin) heated in the electric-loop tool. One end of the string is secured by an alligator clamp attached to a swivel-jointed positioner, the other manipulated by the operator's fingers, as shown in c. The electric-loop tool, with pin fitted to it and heated by the coiled loop, is held in a positioner designed for it. c. A view of an electric-loop tool fitted with a pin (mandrel) for twisting plastic coils, but in this case the tool positioner is designed to be attached to an inner wall of the fume hood and pivot against the wall when not needed. d. Small stepped mandrel clamped to the heating tool, for use in bending plastic to a wide variety of curved shapes. e. View of stepped mandrel, here attached to an alligator clamp and ready for use.

The setup shown here can be used inside the fume hood, but it takes up considerable space and must be removed to make room for routine operations when not in use. A more compact and convenient setup is that shown in Fig. 16, c. Here a dedicated tool is securely inserted into a special holder attached to the lower section of the inner front wall (face) of the hood, the mounting bracket fitted with a swivel joint that permits one to swing the tool outward into a working position when required but as easily shift it back to its unobtrusive storage position. The work-holding clip (a small alligator clip with jaws ground to improve the grip) is mounted on a flexible length of heavy copper wire (insulated, 12-gauge) that permits the necessary adjustment of its position for a wide range of tasks, but it is so mounted on the main tool holder that it is well out of the operator's way when the heating tool is not needed. Small thumb screws or knobs (machine screws glued with epoxy cement into the caps from toothpaste tubes are quite effective) enable one to make all necessary adjustments to the holders and tools, the cork-encased handle of the electric-loop tool, itself held by friction in the brass ring of its holder, permitting easy removal when desired, a particular advantage of this arrangement.

These tools are satisfactory when small and relatively tight curves, loops, and coils must be formed, but when larger, loose or more open ones are required, mandrels used in the setup shown in Fig. 17, a are more efficient. Too large to fit onto the head of the electric-loop tool, these mandrels—easily and quickly interchanged—are mounted on a special holder in such a way that they can be heated quickly by inserting the heated tool into or below them and than as easily cooled either by removing the tool or turning off the heat. The facility shown here is sufficiently small and compact to be used inside the fume hood, inasmuch as the mandrel-holding component attaches to the inside rear wall of the hood and can be swung upwards for storage, away from the work area, the heat source no additional strain on the work space, since it is the electric-loop tool (in its usual cradle) normally used in the work area for general operations. In use, the mandrel holder is swung down into position, a suitable mandrel is attached to it and the loop of the electric-loop tool, in its cradle, is thrust either into the mandrel (if this is sufficiently large) or under it. The range of curvatures producible through the use of quickly interchanged mandrels comfortably encompasses most required in the miniature sculptural pieces. Mandrels are easily made, from thin-walled brass tubing (hardware store or hobby shops) into which is press-fitted a rod that can be inserted into a hole drilled for it in the holder (and held

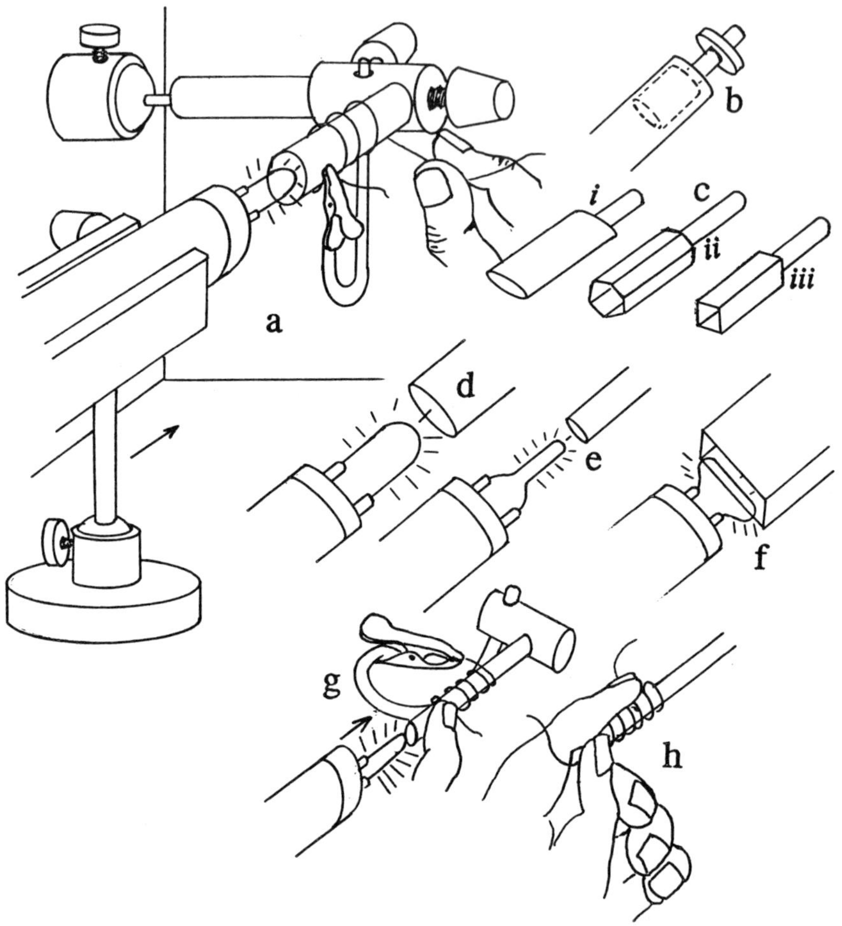

Fig. 17. The use of heated mandrels in shaping miniature sculptural components. a. Twisting a plastic coil around a mandrel attached to a special swivel-joint holder affixed to an inner wall of the fume hood and heated by the electric-loop tool held in its own movable positioner. One end of the coil is secured by an alligator clamp on a flexible arm attached to the mandrel holder. b. Detail of mandrel construction: tubular mandrel (brass) press-fitted to a turned heat-resistant shaft fitted with a Teflon washer for further thermal insulation. c. Mandrels of different shapes, used interchangeably in the holder shown in *a*. d-f. Mandrels of various shapes, showing how the shape of the electric loop is modified to fit each for most efficient heating. g. Detail of coiling procedure, showing the use of the thumb to secure the free end of the coil preparatory to ((h) removing mandrel and coil for cooling (by dowsing in a small water bath) to fix the form and prevent uncoiling.

in place with a thumb screw made from a toothpaste-tube cap, as described earlier). The rod should, preferably, be turned from a heat-resistant (i.e., non-conducting) substance, mine salvaged from an asbestos-impregnated thermoplastic (Bakelite) component of a discarded piece of electrical-engineering laboratory apparatus.Tubing is available in various sizes and shapes, so one should have no difficulty in acquiring the necessary stock. Mandrels smaller in diameter than the loop itself (though this, too, may be reshaped to smaller dimensions if desired, the wire being quite malleable) are made to accommodate it by cutting away the lower half of the working section, most easily done with a small carborundum disk used in the drill press. The addition of a holder for the alligator clip, which attaches to the head of the mandrel-support arm, permits one end of the work piece to be held securely, and it frees both hands for the shaping operation. The arm for the clip being flexible (a short length of heavy-gauge wire), one can easily position the clamped end of the work piece to suit the material and task at hand.

In use, the electric-loop tool is easily slid along its cradle until the correct position is reached to heat the mandrel most efficiently, or the cradle itself may be moved into an appropriate position, the tool clamped securely in it in advance.

The larger mandrel-heating tool shown in Fig. 18 was designed before the one just described and works well in handling a wide range of stock sizes, but it has the disadvantage of requiring a relatively long warm-up time and is, therefore, more useful when large quantities of material must be worked at one session. The heating unit for the mandrels consists of a small electric curling iron, slightly modified to permit one to attach a heavy brass turret, turned, drilled, and tapped to accept mandrels of different sizes and shapes. The mandrels should be attached near the front edge of the turret, leaving the rest for use itself as a mandrel of large diameter. By turning one or more shoulders on the turret, a still greater range of useful mandrel sizes is provided. The plastic tip of the curling iron is replaced by a brass adapter which accepts additional tips when these are required. The tool I use is attached to one side of the fume hood in a conveniently accessible position.

Once a piece of plastic has been heated and shaped by wrapping it around a mandrel of suitable size, it must be held in this form until it cools. Inasmuch as it is not feasible to cool the mandrels individually (another of the drawbacks of this tool's design), a method must be devised for removing the formed piece for separate cooling without having it lose its shape. This can be accomplished (Fig. 18, b, c) by ensheathing it in a paper or plastic sleeve and removing sleeve and formed piece as a unit, cooling this rapidly thereafter, the sleeve assuring the retention of the plastic piece's form.

An alternative procedure is to wrap a wire armature in the form of the desired coil around an unheated mandrel of appropriate size and shape, then coat this with HDPE (using the electric-loop tool to apply it and smooth away the tool marks it leaves) or paint it with acrylics, as appropriate to the design.

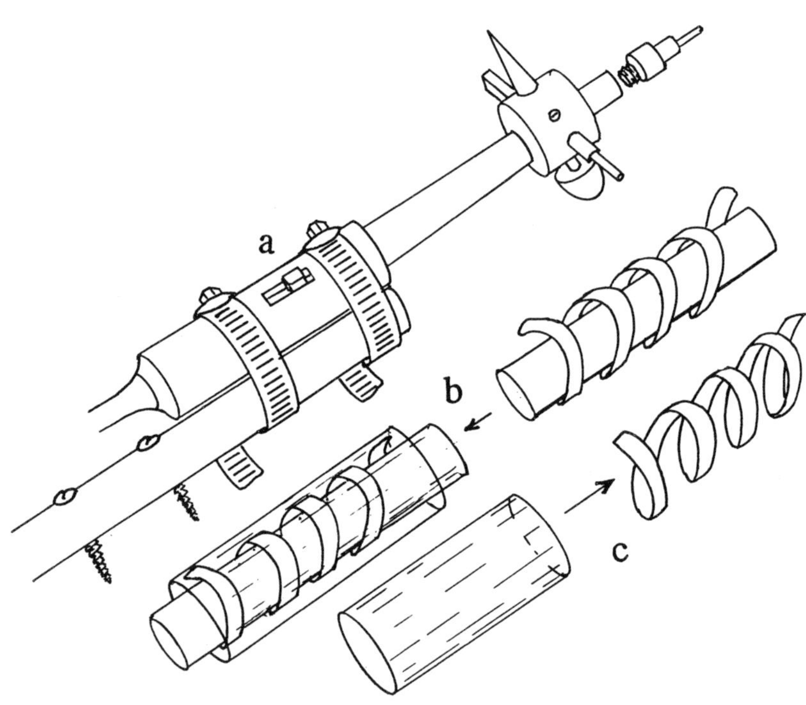

Fig. 18. Forming plastic coils with heated mandrels. a. Turret mandrel for use in forming relatively large components of sculptural pieces. The turret (turned brass), fitted with interchangeable mandrels, is heated by a modified hair curler attached to an outside wall of the fume hood, and a small mandrel screws into the tip of the curler. b-c. Procedure for forming plastic coil and preventing uncoiling during the cooling process. Mandrel and shaped coil are temporarily encased in a closely-fitting sheath of paper or clear plastic until both coil and mandrel cool, after which the sheath is slipped off and the cooled coil will not lose its shape.

AN ODE TO MANDRELS

A mandrel-less shop is a shop without bliss.
If you've any doubt, I'll explain it like this:
Mandrels are helpful in holding one's work;
They simplify tasks not easy to shirk.

Often they serve as a cheap extra hand
When the force at work is undermanned.
They help to prevent the piece that they hold
From failing to do just what it is told.

They stop it from wobbling or flopping about,
And greatly increase the craftsman's clout
As he applies a tool with precision and grace
To a piece that the mandrel helps keep in place.

A well-designed mandrel is a joy to use,
A treasured tool one dare not abuse.
In size and shape it has to be right,
Otherwise you're left in an awkward plight

And wish you had stayed at the piano instead
Of the lathe or mill to worry your head.
But since you have chosen the banausic approach,
The subject of mandrels is one you must broach.

A mandrel's importance we all recognize;
Our love for such tools we'll never disguise,
So give us more mandrels, we can't have too many!
Far better that way than not to have any.

8

EXTRUDING PLASTIC STOCK

WHILE STRIPS of plastic cut from scrap stock with a heavy pair of scissors or a paper cutter may suffice in building up one's sculptural pieces, I have found it most helpful to be able to supplement these with rod, strings, and threads extruded by feeding strips into a glue gun, mounted for the purpose on the work bench. In the case of hot-melt glue, extrusion is, without question, the most efficient way of producing stock material in diameters suitable for miniature sculptural work. Extrusion to form the polyethylene and polypropylene stocks has the special advantage of enabling one easily to produce material in the size most appropriate for practically any task to be performed in making a sculptural piece, to say nothing of the ease with which the mixing (through coextrusion) of various materials to achieve desired effects of color patterns, textural features, and working properties (such as ductility, for example) may be accomplished using the process.

An efficient and inexpensive setup for extruding these materials is shown in Fig. 19. To extrude most of the types of plastic used in miniature sculptural projects considered in this book, one need only attach a glue gun (of which several types are readily available in stores today) to the work bench, operate it through a variable transformer to achieve proper temperature control for the particular plastic being worked, and ram strips of the material into the heating chamber with a short length of wooden dowelling. Pressure applied to the dowel forces the heat-softened plastic out of the nozzle, and it is then a simple matter to pull (with forceps, pliers, or a wooden stick) the extrudate into rods, strings, or threads, depending largely upon the temperature of the heating chamber and nozzle and the speed at which one pulls. To prevent the extrudate from adhering to the outside of the nozzle and complicate the pulling process, pull the first bit clear of the nozzle and then, as pulling proceeds, continue to pull gently, directing the extruded material away from the gun for the distance of one's reach, wiping the cooling length from the nozzle and restarting a new length, the process repeated until the supply of molten stock (or the operator) is exhausted.

A more satisfactory procedure is to tilt the extruder downward so that the molten plastic falls onto the moving belt of the take-off device shown in Fig. 19, g. With this setup, the extrusion process is made much more efficient, the molten plastic not now tending to curl around the nozzle and adhere to it as it does when operated in the horizontal position and without a takeoff device. Instead, it falls directly onto the moving belt and is carried away during the cooling process. A homemade belt, mounted on metal or hardwood rollers in a length of aluminum channel stock, is moved by means of a handle at one end of the channel. The extruded plastic cools rapidly as it is carried along by the belt and passes off the far end of the device, falling toward the floor in a continuous length, after which it can readily be cut into suitable lengths for use or storage. The belt is driven by friction with a roller at the far end of the channel and runs freely on one at the other, a third one (in the center of the run) serving as an idler which can be adjusted to increase the friction on the drive system and prevent belt slippage. The position of the middle roller in its vertical mounting slots can be adjusted by means of a slotted plate over each end of its shaft. The belt can be made of heavy wrapping paper, the ends spliced and glued to form a continuous loop. An overlay of electrical tape provides a good working surface, and its black color gives good contrast with the ivory and white plastics used for many of the sculptural pieces. Two simple aluminum brackets are used to clamp the take-off device in position against the face of the work bench, in readiness for an extrusion session, but the quantity of extrudate producible in a half-hour session is such that these are infrequent.(*Such long sessions will probably shorten the gun's life, it not being designed for extruding these materials. **Important:** Note the various safety precautions and qualifying remarks given at the end of this chapter and elsewhere in the book.*)

The simple arrangement shown in Fig. 19, a, the gun attached to a metal upright by means of adjustable hose clamps, enables one to screw the mount to the work bench, but the upright can be screwed instead to a length of channel aluminum fitted with tightening screw with sliding handle (Fig.19, c) and then the whole attached temporarily to the work bench.

I have found it helpful to have one or two supplementary nozzles to expand the range of diameters one can produce. With smaller nozzle diameters than those supplied with some guns, one automatically reduces the initial diameter of the extrudate and can then reduce it further by controlling the variables already mentioned. Nozzles may be turned on the lathe and tapped to fit in place of the one supplied with the gun or merely added to the existing nozzle by appropriate tapping.

One should exercise care in selecting the plastic to be extruded, the first considerations being appearance and working properties. The pulling

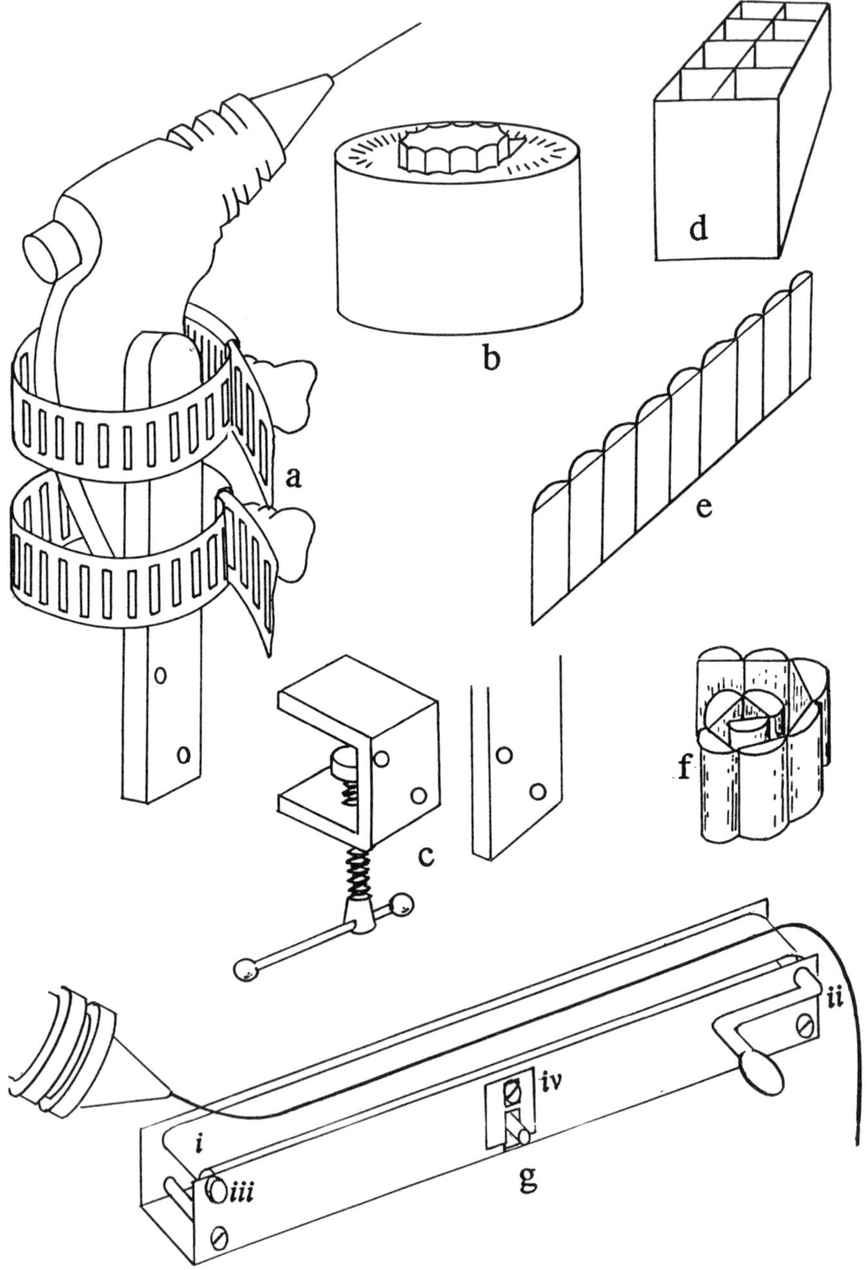

Fig. 19

properties—critical in the extrusion process—vary considerably, so one should test these before beginning. This is easily done, merely by heating a small quantity to the melting point and attempting to pull it out to a fine thread with a needle or toothpick. Alternatively, allow some of the molten material to drip from the main mass to see if a fine thread of plastic joins the dripping mass to the upper one as they part. If threads are easily produced in this way, the plastic should extrude nicely and subsequently respond well to the electric-loop tool when being worked into a sculptural composition. If, by contrast, however, the material, when heated, merely forms an obstinate irregular mass of small rounded hummocks that cannot easily be pulled out into fine filaments or will not easily drip away from the parent mass, then it should be rejected, because the likelihood is that it will be difficult to extrude in the way one would like and will not respond cooperatively to one's subsequent efforts at shaping it with the electric-loop tool and needle or forceps.

Important considerations when selecting plastic material are its color and its light-transmitting properties. I generally prefer either white or ivory, but in either case and for most purposes the material should be opaque. If transparent or translucent material is required for some special task, one can use extruded stock of hot-melt glue (the transparent, rather than the opaque type), although in some cases low-density polyethylene (without fillers) might be suitable. If the plastic one wishes to use lacks the drawing properties required for proper extrusion, the addition of hot-melt glue to the stock material as it is fed into the heating chamber of the glue gun may correct the deficiency, but as a general rule it is a simple matter to select with care (and with a simple test of drawing properties in advance) a suitable plastic among the food containers normally encountered around the house.

Fig. 19. Extruding plastic stock for miniature sculptural pieces. a. Glue gun ready for use in extrusion. The gun may be attached directly to the work bench, as shown, or, alternatively, it may temporarily be attached by means of a clamp like that shown in *c*. b. Powerstat transformer for use in regulating the temperature of the extruder. c. Homemade clamp for attaching the gun temporarily to the face of the work bench. d. Caddy (for extruded rod stock) made from a small cardboard box to which partitions have been added. e. Polyethylene caddy made with a domestic bag sealer. f. Polyethylene caddy coiled for convenient storage and access. g. Takeoff device for spreading and cooling plastic as it is extruded from the glue gun: *i*. Endless belt in its aluminum channel. *ii*. Drive shaft (right); *iii* Idler (left); *iv*. Adjustable belt-tightening idler (center).

While I have found a glue gun (even the miniature type) quite adequate for my purpose, others with more ambitious programs may wish to aquire more capacious extruders. A larger glue gun is the next logical choice, of course, but, with a little experimentation, a clever artisan should be able to make a simple extruder from workshop scrap, though caution should be exercised in doing so, because the hazards inherent in such gadgeteering could lead to grief.

I must emphasize the fact that the glue guns one buys were *not designed for use with some of the materials suggested in this book nor for the relatively protracted extrusion sessions that this involves.* The likelihood is that with this obvious misuse or misapplication of the tool its working life will be considerably shorter than it might otherwise be. In my experience, however, the expense of such admittedly crude or inefficient improvisation i.e., having occasionally to replace a glue gun, is not an unbearable extravagance. *One should also be aware of the potential hazards involved in having a hot gun attached to the workbench as one proceeds with the extrusion process. Provisions should be made: (1) to remove noxious fumes from the heat-softened plastics so that they are not breathed in by the operator; (2) to prevent burning oneself by accidentally touching or coming into contact with the hot nozzle of the gun or by letting molten plastic come into contact with any part of one's body; (3) to prevent the take-off belt, the workbench, or any flammable material from coming into contact with the hot nozzle of the gun, possibly causing a fire; (4) to deal quickly and effectively with any fire that might conceivably occur during the operation (have fire extinguisher, fire blanket, and /or other emergency aids close at hand. As a special precaution, wear gloves, eye shield, an aspirator, and suitable protective clothing (though I myself am not as fastidious as this).*

As shown in Fig. 19 (d-f), convenient caddies for storing the lengths of extruded stock can be made by gluing together sections of rectangular cardboard packages of the type common about the house (packages for toothpaste and ointment tubes, etc.) or by reworking polyethylene-bag plastic with a domestic bag sealer, mine costing $3.00 at a thrift shop ten years ago (and still quite serviceable).

9
ADAPTATIONS OF SOME CONVENTIONAL PROCEDURES FOR WORKING THERMOPLASTICS

ON OCCASION it might be easier to make some component of a sculptural piece in another material than thermoplastics, make a mold from this pattern, and then produce a replica from this for use in the final piece, yet another reason for familiarizing oneself sufficiently with a wide range of molding, casting, thermoforming, and other procedures for duplicating and replicating objects. With a wide range of such techniques at one's fingertips, one's design horizons are immeasurably extended and permit a variety and diversity not so readily achieved in thermoplastics by the basic sculptural processes of simple carving and modelling. *Bear in mind, however, that applying some of the procedures described here exposes one to the types of hazards with which any experimentalist should be familiar. One should proceed with great caution and take all steps necessary to prevent accidents or injury. Read the precautionary notes and warnings given at various places in this book, and consult some of the extensive safety guides available in the literature of plastics technology before undertaking such procedures.*

Most of the major processes for working and forming thermoplastics may be of use in preparing these materials for incorporation into miniature sculptural pieces, though extrusion is the principal one and may, in fact, be the only one if one prefers to keep technical complication to a minimum. By adding to one's facilities the means for applying the others, however, one can greatly expand the range of achievable effects and, because the size of the finished pieces is so small, the tools for producing them may be simple and inexpensive enough to be made in a modest home workshop.

As already mentioned, one of the more obvious advantages of working at a miniature scale is that it permits the easy and relatively inexpensive application of many sophisticated processes for working both thermoplastics and other of the more usual craft materials that at larger scales would be impracticable, a case in point being the blow forming, blow molding, and vacuum forming of polyethylene and polypropylene. In some cases, these processes may be used to great advantage instead of or in conjunction with other procedures, such as compression molding, drape molding and forming, embossing, and stamping.

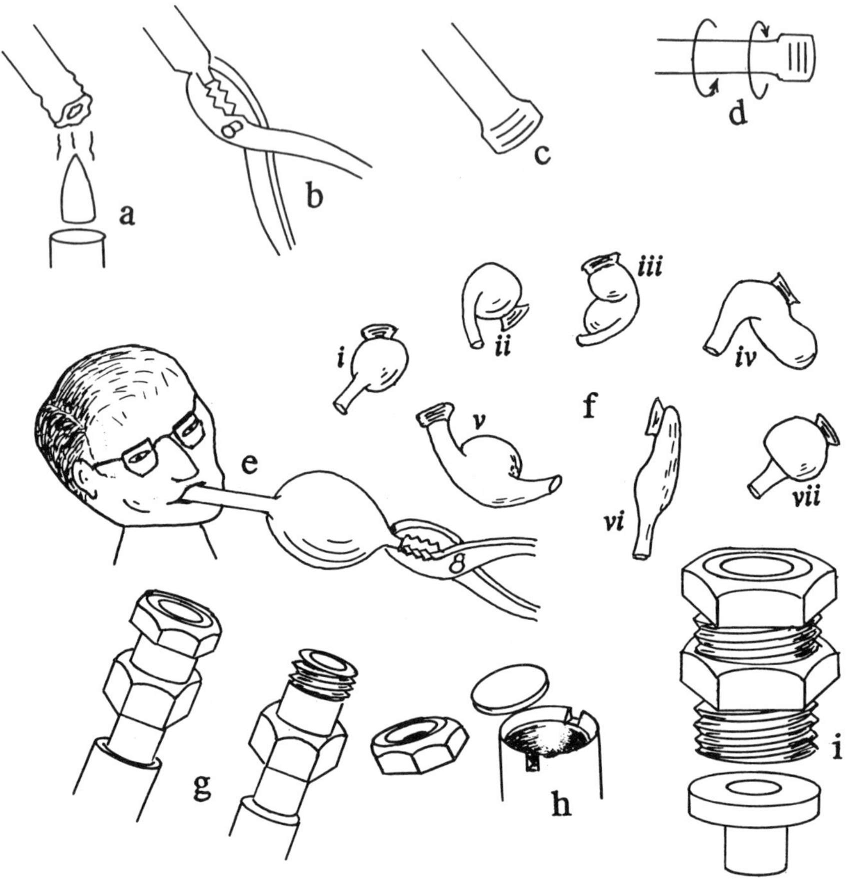

Fig. 20. The blow-forming process and tools. a-e. The process of free-form blowing. a. Melting the free end of a length of polyethylene tubing. b. Closing the tube with pliers. c. The closed tube. d. Rotating the tube while the section to be blown is heat softened. e. Blowing a bubble in the tube (note that pliers must constantly be used to keep the end of the tube sealed during the heating and blowing process. f. Various shapes easily produced by blowing. g. Two views of a simple multipurpose tool used in blowing small objects. A plastic blank is held over the orifice of the brass nut. h. Hollow drill (homemade) used for cutting plastic blanks (diam.: 0.75"). i. Larger blow-forming tool, one that accepts 1.75" blanks. Both blow-forming tools are multipurpose ones, suitable for various plastics-working procedures, including free-form blowing, blow molding, vacuum forming, yoke/plug molding, and combinations of some of these, such as yoke/plug and blow molding.

Blow forming with and without molds

Blow forming, either with or without the use of molds, is one of the techniques that adds considerably to the range of forms usable in thermoplastics sculpture. At the scale being considered here, the equipment required for the process is simple, consisting merely of some stock polyethylene or polypropylene in the form of either sheet or tubing, a pair of pliers, and a small gas burner (laboratory type or propane torch). A length (one or two feet) of low-density polyethylene tubing will be sufficient for preparing a dozen or so small blown objects that might be used in sculptural pieces, as would the sheet plastic salvaged from a single quart-size container of high-density polyethylene.

To blow a bubble in tubing (Fig. 20), one end is first melted and securely sealed by pinching it together with pliers and then allowing it to cool. A section of the sealed tubing is then heat softened and blown by applying pressure at the free end. By blowing carefully and gently, a bubble can be formed without rupturing, and, with a little practice, one can quickly gain skill in producing forms of a size and shape suitable for use in sculptural miniatures. During the heating process, the plastic changes form translucent to transparent, its watery appearance at this latter stage the cue to remove the tubing from the heat source and begin blowing. The tendency is to blow too hard. With practice, one can soon learn to work sufficiently well within the dictates of the variables involved to be able to produce a variety of useful forms by this means, but the scope is limited and is soon likely to lead to the use of molds and other devices for enhancing it.

If one wishes to add color to the blown object, this may be done either by filling or coating the inside wall with molten wax or by painting the outside (with acrylic paint, for example). To fill a blown cavity or merely to line it with wax, one pipettes molten wax (from the tip of a finely-drawn pipette that can be inserted into the tubular opening of the piece) into it while the body of the object is immersed in cold water, this to prevent melting its exceedingly thin and delicate wall. *Avoid overheating the wax before pipetting it, and practice on unimportant pieces before attempting it on valuable ones.* Filling must be done with great care, the object gradually submerged as this proceeds. *When heating the wax to be pipetted, special care should be exercised to avoid excess heat, fire, and spilling (an electric hotplate, thermostatically controlled, here preferable to an open flame, the temperature just sufficient to melt the wax, not superheat it).*

Unfilled bubbles or other hollow forms produced by blowing are usually transparent or nearly so, meaning that objects placed inside them can be seen from the outside and become a part of the visual whole. One can introduce objects through the tubular orifice of the blown piece or the wall can be pierced by careful melting (with the electric-loop or heated needle tip of the tool, but such procedures require delicate, precise use of the tools and a steady hand.

The two tools shown in Figs. 20, g and i are multipurpose ones for use in thermoforming blank disks cut from sheet-plastic stock. The smaller of the two (Fig. 20, g) was designed for use in blow forming (with or without molds), but experimentation soon revealed its value in other processes as well, particularly in vacuum forming and in yoke-plug molding. Produced on a lathe from gas fittings, its construction required only the following steps:

1. boring the threaded end slightly, to provide maximum space for the working chamber.
2. removing all but two of the threads from the outside (male element) of the fitting.
3. reducing the height of the hexagonal nut to about 0.25", leaving only two threads to engage the two left on the male element of the fitting.
4. boring a shallow groove at the inner end of the threaded section of the nut to permit it to hold the plastic blank securely (and air tight) when hand tightened.
5. removing enough metal from the outer face of the nut to reduce the distance between it and the plastic blank to approximately 1 mm., this to facilitate rapid, uniform heating of the plastic disk.
6. adding (by soldering) a simple turned adapter for attaching a length of rubber tubing (for use in applying pressure or suction with the operator's own lungs).

The larger of the two tools (Fig. 20, i) was also made from a brass plumbing fixture modified along lines similar to those just listed, the upper threaded ring turned inside to accept the plastic blank and, when required, a small mold. With its more capacious working chamber, this tool enables one to produce larger pieces than the first one and is, thus, more versatile.

To use either of them, a plastic disk (cut from a yoghurt container, for example) is inserted against the threaded face of the male element, the nut is screwed securely in place (to form an airtight seal), and sufficient heat is applied (by holding it briefly over a small gas flame) to permit the operator to produce the desired form (by blowing or sucking on the attached length of tubing). A little experimentation on scrap pieces will enable one to determine the size of flame, the appropriate distance from flame to disk, and other controls needed to gauge (simply by counting) the necessary heating time.

The simplest thermoforming operations to be performed with the tools require no molds or patterns, but the range of shapes thus produced is limited largely to bubbles or dome-like structures (one advantage of these over those produced by blowing tubing, however, being that their base or back side is open, permitting easier filling or additions to the inside). The size of these may be varied by reducing the area of the blank heated, the amount of pressure or suction applied, and the speed at which this is done.

A powerful approach to achieving diversity of form, structure, and texture is to use various supplementary tools, such as molds, dies, stamps, plugs and yokes over or into which the heat-softened plastic can be forced through pressure or suction. If such a supplementary tool must be placed inside the tool chamber, however, provision must be made for positioning it properly and holding it there during the forming operation. This is complicated by the fact that the tool must be inverted over the burner during the heating process, this inversion, of course, subjecting any free component to gravitational force and displacement unless steps are taken to prevent it. In some cases it may not matter if the accessory piece falls onto the inner face of the plastic disk during the heating and forming process, but in others the piece must be kept fixed or properly aligned, in which case a perforated base plate (a stiff disk of copper or brass screening is good here) must be inserted so that the component can be attached to it (soldered, glued, etc.), the whole held in position by spacers and retaining rings cut from brass or aluminum tubing of appropriate diameter and wall thickness. With a set of such spacers and rings on hand, it is a simple matter to position a mold, for example, properly within the working chamber and prevent if from moving about during the heating and forming process. Some of the difficulties here being considered may be eliminated by applying heat from above to the blank of plastic material to be formed (a small electric plate, controlled with a variable transformer and mounted on a swivelling arm, the forming tool itself held rigidly in a clamp attached to the common vertical support).

A varied stock of retaining screens, spacers, and rings equips one to use an extensive range of molds, forms, yokes, plugs, and patterns within the working chamber and to perform with this simple equipment a wide range of thermoforming operations on a scale appropriate to miniature sculpture.

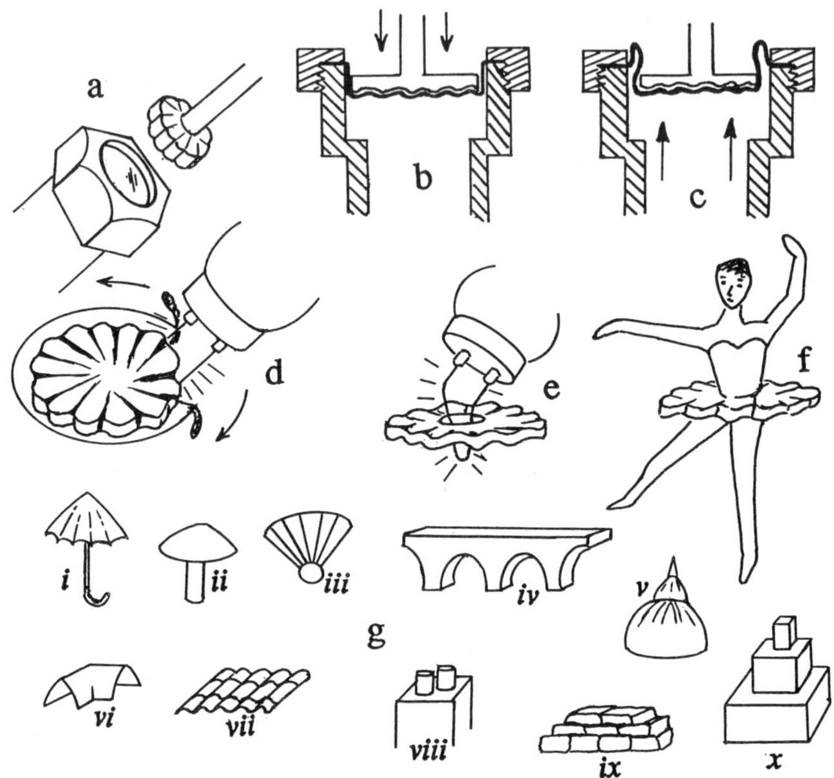

Fig. 21. Blow-molding procedures for producing small components of min-iature sculptural pieces. a. Male mold and multipurpose tool ready for use in preparing a tu tu for a ballet dancer (f). The plastic blank, ready for heat softening, is held inside the tool and (b) the male mold (plug) is pressed into the heat-softened blank. c. Pressure is now applied (by blowing into the tube attached to the tool) and the formed piece is allowed to cool. d. Excess plastic is trimmed from the piece through the use of the electric-loop tool and (e) a hole is melted in the center of the tu tu for inserting the figure and completing the piece. f. Ballet dancer with blow-molded tu tu. g. Various forms easily produced in a similar way: i. umbrella, ii. toadstool, iii. fan; iv. viaduct; v. cathedral dome; vi. groined roof or ceiling; vii. tile roof; viii. chimney; ix. stone wall; x. skyscraper. Each of these pieces was formed from a 0.75" blank.

Generally speaking, it is easier to use such accessories outside the multipurpose tool and with blow-forming procedures. With vacuum forming, for example, the mold must be placed inside the tool chamber and there positioned with suitable retaining rings or spacers, but with blow forming the mold is simply hand held outside the tool. One effective way of assuring good registration for intricate detail is to press a male mold element well into the heat-softened plastic blank (Fig. 21, b) and then blow the plastic vigorously against it (Fig. 21, c). Some of the forms easily produced in this way are shown in Fig. 21, g, various others as components of the finished piece illustrated in Plate 6, k.

One application of the blow-forming process is in making molds from heads and faces or other structures on pre-existing figures. From such molds—so easily made by this process—one can cast (in epoxy or polyester resins, for example) replicas of components suitable for use in miniature sculptures. Patterns of various types of structures, from mountains to skyscrapers, can be made in plaster or wood and then used to produce blow-formed molds from which additional cast replicas can be made. By drape molding (blow forming) heat-softened plastic over a suitable wire armature, umbrellas, fans, and membranous structures like the wings of butterflies and other insects are easily produced.

The special problem for the miniaturist of producing realistic pleats and folds in garments can sometimes be resolved successfully through the use of blow-molding. The simplest approach is to blow a thin sheet of heat-softened plastic over an appropriate male pattern (in boxwood, acrylic, or metal), this being easier to make than a female one (or mold). The groove in the pattern should first be made with a fine (jeweller's) saw and then rounded with a file. After the plastic is blown around the pattern, excess material is trimmed away, and, since the blown form is thin and easily melted with the electric-loop tool, its cavity should be filled with plaster for support. If the blown film is too thin, use heavier sheet stock in the blowing process. If, too, the pattern is inserted loosely within a tubular sheath, better registration and reproduction of detail will be achieved, since more of the film is concentrated on the pattern itself and less of it dissipated by unrestrained expansion. The base of the pattern should fit sufficiently loosely within the sheath to permit the blown film to press securely over it and provide a firm surface (when cooled) for trimming.

If one prefers to use a female pattern, or mold, for producing pleated garments, this must initially be hollow (in order to use a jeweller's saw and pattern files to make grooves in it). This mold can be filled by pressing molten plastic in it to produce a solid casting, or a delicate film replica can be made through blow molding or vacuum forming, using the larger of the two multipurpose tools, the principal difficulty in either case being to obtain good registration and reproduction of detail. In the case of vacuum forming, the center of the mold should be plugged before use to prevent plastic from being

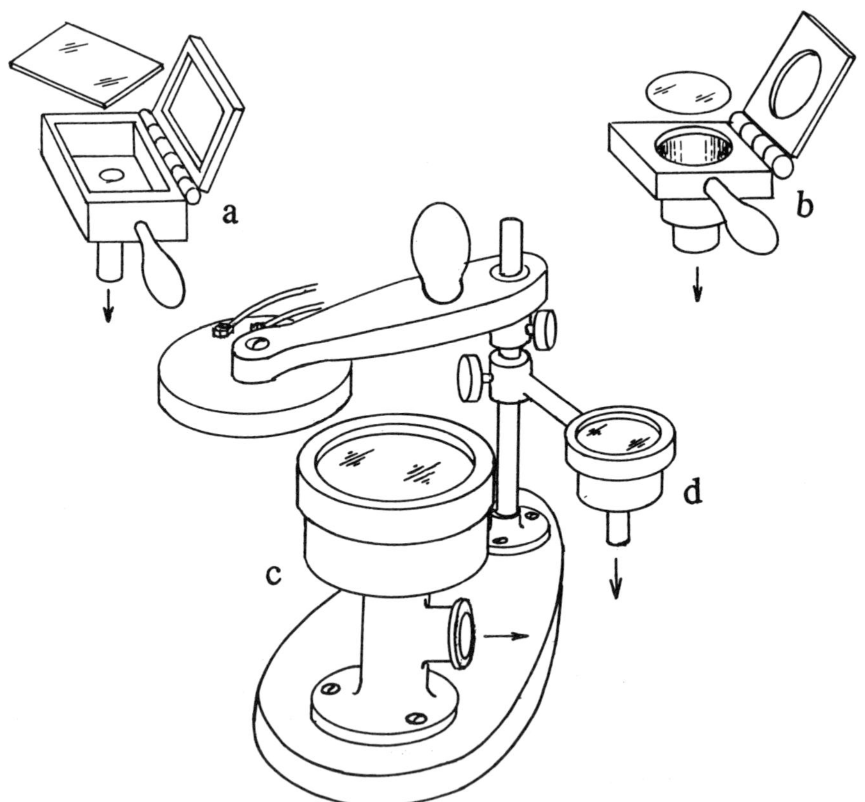

Fig. 22. Tools for vacuum-forming components of miniature sculptural pieces. a. A small former for use with suction supplied by the operator through a rubber tube, heat from a miniature burner (or even a candle). This tool uses rectangular blanks measuring 1"x 2". b. A similar small, hand-held, mouth-operated former, designed to accept blanks 0.75" in diameter. c. Vacuum former designed to take 4" blanks, use vacuum supplied by a domestic vacuum cleaner and heat from a 6" ceramic heating plate. The heating unit swivels around the upright that supports it and can be used with the smaller (2" blanks) mouth-operated tool (d) attached to the same support.

sucked out through the opening at the bottom of the mold and even torn loose. The plug should be so designed that it supports the plastic at the bottom of the mold but does not impede the flow of plastic all around it. It can be made to fit flush against the base of the mold (held in place by epoxy cement or cyanoacrylate glue) or it can be made to fit against a shoulder turned for it inside the mold.

Vacuum forming

Vacuum forming being such a useful component of one's facilities, it may well be worth the effort of making a larger tool for the task than the multipurpose tool shown in Fig. 20, i. Constructed as shown in Fig. 22, c, the larger tool has a 6-inch heating plate and a fitting for attaching a domestic vacuum cleaner (as the vacuum source); such a tool is easily assembled from pipe fittings and scrap bits and pieces. Its larger capacity is not essential to the production of miniature sculptural pieces, the two multipurpose tools described earlier being adequate for most needs, but such a tool can be useful in many ways, from making molds to be used in replicating components to producing stocks of forms that might be required in quantity (tile roofs, brick or stone walls, etc.). The potential uses are legion.

Yoke/plug molding

Another valuable procedure for working thermoplastics for inclusion in sculptural miniatures is that of using a matched yoke and plug to shape the heat-softened plastic. The male mold element (the plug) is simply used to force the heat-softened plastic into (and sometimes through) the female element (the yoke), the form of the male imparted to the plastic as it passes through the yoke, the latter exerting the restraining influence essential to the proper registration and replication of the form of the plug. Small components may be produced with the multipurpose tools shown in Fig. 20, g, i, if one of the two mold elements (usually the yoke) can be positioned securely within the tool chamber and the other (the plug) pressed home from the outside when the plastic has been properly softened. An advantage here is that additional pressure—from blowing— can be applied to achieve good registration between the plastic and the plug. An alternative method (and one that allows more scope and freedom and often is easier and simpler to use) is to clamp a disk of sheet stock between two metal rings (Fig. 23, a), the plug

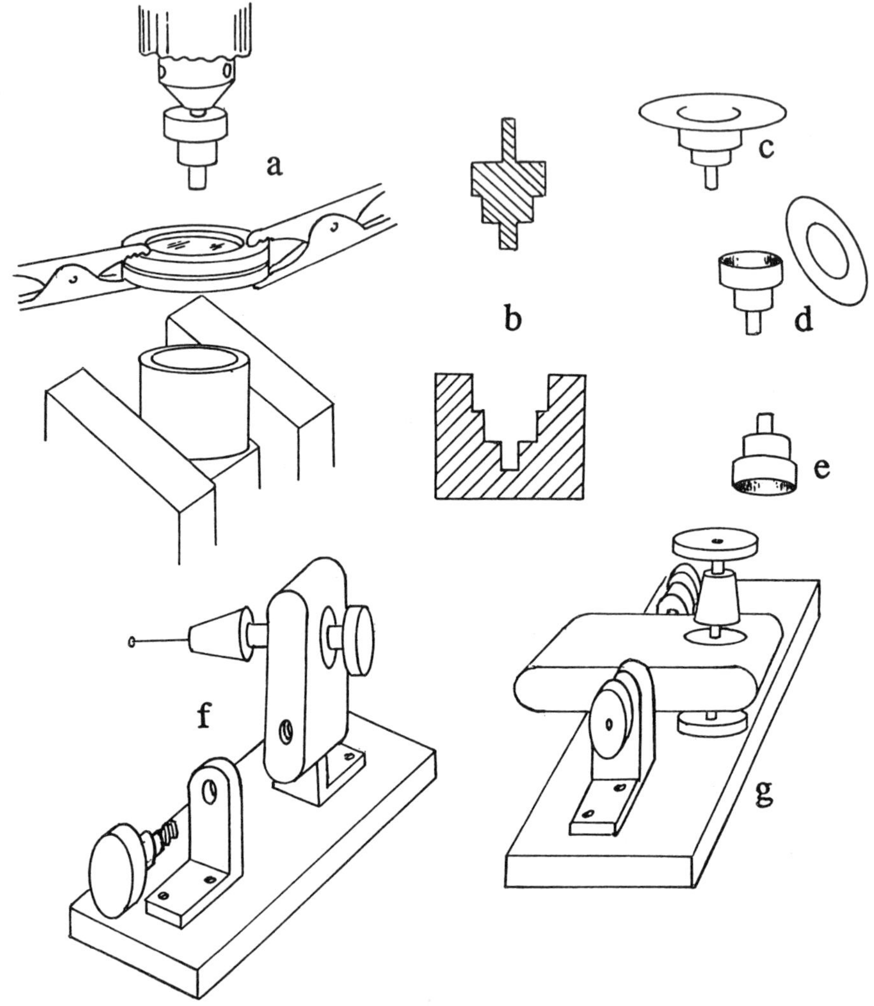

Fig. 23. Molding and turning components for miniature sculptural pieces. a. Tooling for the yoke/plug molding process. The male mold element (plug) is held in the drill press, the female (yoke) in a drill-press vise as the plug is forced to press part of a heated blank of plastic into or through the yoke. b. Cross sections of the plug (*i*, above) and yoke (*ii*, below). c. The formed piece, prior to trimming. d. Excess plastic trimmed away. e. Finished piece, ready for incorporation into a sculptural piece. f. Manually-powered lath (partly exploded) with straight pin chucked as a mandrel upon which plastic will be melted for further buildup and shaping (see Fig. 24, a). g. Lathe rotated to place its shaft in the vertical position for use as a potter's wheel.

held in the chuck of the drill press, the yoke in the vise on the press table. The clamped plastic is then heated over a gas flame (laboratory burner or propane torch) and transferred to the forming area, the plug simply pressed down to carry it into the yoke the required distance (set in advance with the quill stop of the press). Generally speaking, the forms to be produced by this procedure are simple ones, the problem of undercuts the principal limitation to be considered, but with a little experimentation one can expand greatly the range of applications to be found for it. By using a stepped plug, for example, one can produce the structure shown in Fig. 23, e. As in most molding operations, a slight taper on the sides of the plug facilitates extraction after the mold is closed, and a mold-release agent (silicone spray, for example) can be most valuable. Boxwood is an excellent material for making both yoke and plug.

Turning and throwing

A helpful addition to one's tool kit is a simple lathe, either manually or electrically operated, for use in turning thermoplastics. Used in conjunction with the electric-loop tool, such a device enables one to produce a range of radially-symmetrical objects not easily formed freehand either by carving or modelling. An essential feature of such a lathe, of course, is that it be small enough for efficient operation inside the fume hood. The electric one shown in Fig. 24, a, for example, is only 4.5" tall (including the base).

A simple hand-operated lathe like the wire-twisting tool shown in Fig. 9, a, consists of but a small pin vise (Starrett no.162 c , for example), a short length of plastic tubing (Tygon), which serves as a sleeve bearing, a modified clothespin (spring type), and (not shown here, this one being mounted in a vise instead, for use in twisting wire armatures) a heavy metal base (to give the stability essential to efficient operation). The addition of a knob or handle to the free end of the vise, though not essential, makes the tool easier to use. Such a rudimentary lathe as this will enable one to produce a modestly wide range of forms. A slightly more sophisticated version is shown in Fig. 23, f, this version easily rotated to serve as a potter's wheel as well (Fig. 23, g). Using the electric-loop tool for the operation, the plastic to be formed is melted onto a straight pin (chucked in the pin vise), and, while the lathe is slowly rotated, more plastic is melted and made to flow onto the mass being built up on the pin. Once well attached to the pin and built up to the volume required, the mass can be shaped with the electric-loop tool and then smoothed

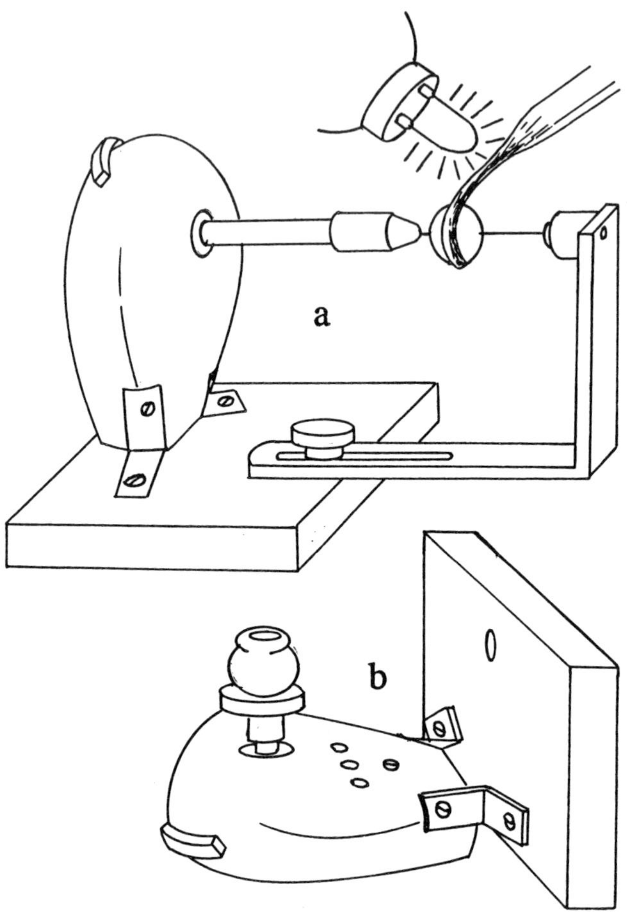

Fig. 24. Electric lathe and potter's wheel for working thermoplastics. a. Electrically powered lathe in use. A straight pin serves as the spindle or mandrel on which plastic is being built up, using the electric-loop tool to melt additional material onto the spherical mass. The tailstock consists of a Teflon plug set in a hardwood housing attached to an adjustable aluminum support. In use, the tip of the pin is pressed into the Teflon plug, which makes a heat-resistant chucking device. b. The lathe turned onto its back, serving in this position (shaft vertical) as a miniature potter's wheel, a pot shown attached to the wheel plate after being "thrown" as shown in Fig. 25, e-h. For this conversion, the tailstock assembly is not needed, so it is removed, although with but slight modification (lengthening the adjustment slot on the support) it could serve as a tool rest during the throwing process. Height of lathe: 4.5"; length: 4.25- 5.5" (tailstock contracted or extended).

to remove tool marks. Such a lathe is particularly useful in making turnings in great variety (Plate 14), including perfect spheres and ovals difficult to produce freehand. The finished piece may be left on the pin (trimmed to the desired length) and attached directly to the developing sculptural piece, or the pin may be removed (mild heating of the pin sometimes simplifies its extraction) and the formed component fused in place on the sculpture.

An electrically-operated lathe is, needless to say, considerably more convenient and easy to use than a manually-operated one and is not difficult to construct, the basic need being for a small motor capable of operation with a speed-control device (preferably with a foot-switch, to free both hands for use with the lathe and electric-loop tool). The lathe shown in Fig. 24, a consists essentially of an outmoded model of a Keufel and Esser erasing machine, this fitted with: (1) a chuck cut from a bacteriological innoculation loop-holding tool; (2) a heavy steel base (3"x3"x0.50"); (3) a simple aluminum-strap frame for the tailstock (a plug of Teflon inserted in a turned wooden sheath, the latter screwed to the aluminum frame), the position of the frame adjustable, as shown in the figure; (4) a variable transformer (Powerstat, model 10B); and (5) a homemade foot switch (microswitch in a wooden housing). This miniature lathe, only 4.5" tall, can be operated efficiently within the confines of the fume hood, and it can easily be removed when not needed. As with the manually-operated one, this one uses a straight pin as a mandrel, the head of the pin removed and the sharpened tip merely pressed into the Teflon tailstock, the decapitated end, of course, chucked in the headstock.

This lathe is easily converted for use as a potter's wheel (Fig. 24, g) merely by removing the tailstock assembly and turning the motor onto its back. Some forms (such as pots and urns, for example) are more easily produced when the work rotates around a vertical axis (as on a potter's wheel) rather than around the horizontal one of the lathe. The turntable of the simpler, hand-turned potter's wheel shown in Fig. 23, g is merely a disk of mounting board pierced by a straight pin, the latter chucked in the jaws of the pin vise, the pin cut to a length that keeps the disk as close to the chuck as possible (this to minimize wobble). In use, plastic to be worked with the tool is melted onto the disk to form a secure bond (a bit of hot-melt glue can help here) and additional plastic is melted in place (using the electric-loop tool) to give the required volume, after which the heated tool is used in conjunction with a small needle tool to shape the mass to the form desired. Many of the operations typically employed by conventional potters can be applied to working thermoplastics on either the manually-operated lathe or the electric one. A little practice and experimentation will soon enable one to devise the best tools for shaping the molten mass and to develope the technical finesse

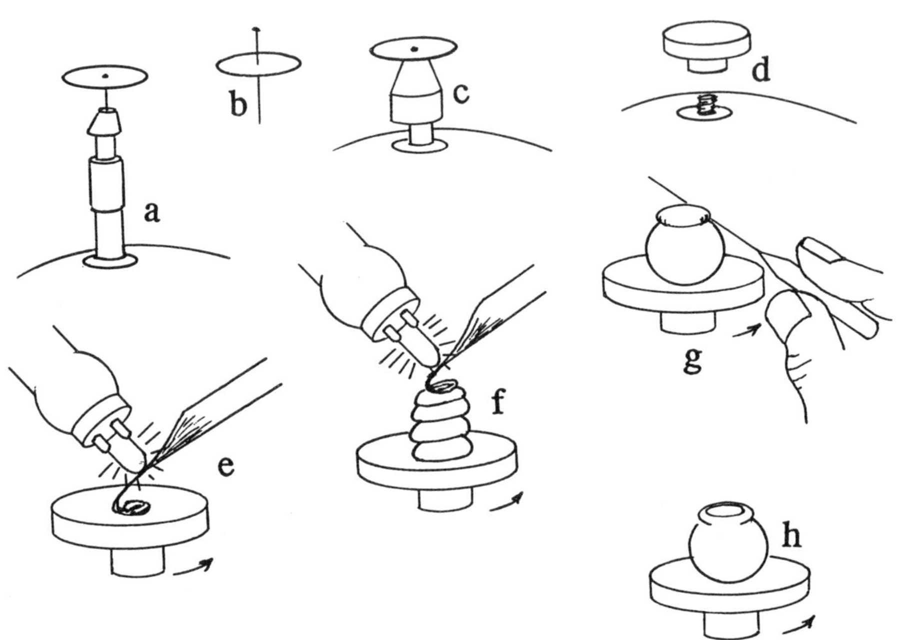

Fig. 25. Potter's wheel and the process of throwing small thermoplastic objects. a. The drive shaft, chuck (pin vise) and pin-mounted wheel for the electric potter's wheel (shown in Fig. 24, b). b. Cardboard wheel mounted on a straight pin. c. Shorter pin vise (replacing the longer one shown at *a*) used as a chucking device for the pin-mounted wheel, this resulting in truer rotation (less wobble) than with the longer vise. d. Metal wheel (brass) turned to screw directly to the threaded shaft of the motor (from an electric eraser), the most efficient and accurate of the designs. e-h. The throwing process: e. Early stage in building up a mass of plastic to be thrown; f. Final stage in the buildup, prior to the initial smoothing out of the coil and preliminary shaping, all with the electric-loop tool (not shown); g. Late stage in the shaping process, the plastic heat-softened (electric-loop tool) in the area to be shaped, a mounted needle or pin used to impart the desired details of contour. h. A finished urn on the wheel, ready for removal (electric-loop tool used to heat the wheel sufficiently to free the piece from it).

required in achieving the goals one sets for this phase of the sculptural effort.

In both the lathe and the potter's wheel—whether manual or electric—the precise control of the speed of rotation is critical, as is the control of the amount of heat applied to the plastic being worked. A foot switch to activate or stop heating (and the blower to remove fumes) is essential, too, as is a separate one to control (on/off) spindle rotation; the intensity of heat and the speed of rotation are set manually in advance on the variable transformers dedicated to each. The foot switches free both hands for these and other operations essential to the use of the tools themselves.

The electric-loop tool must be held at the proper distance from the plastic to assure most efficient heating (sufficient to induce adequate, but nicely controllable, flow), and the loop itself must be shaped and reshaped as needed to enable one to control the application of heat in the most effective manner. Thus, for example, when one must open up the interior of an urn on the potter's wheel (Fig. 24, b) the width of the loop must be reduced sufficiently for it to pass into the constricted aperture without enlarging the latter in the process. Often the loop must be bent to correspond to the curvature of a surface being worked.

Of the two operations, turning on the lathe and throwing on the potter's wheel, the latter is probably the more difficult, owing in part to the fact that a mandrel can more readily be used on the lathe than with the open or hollow objects one more frequently produces on the wheel. Solid objects, particularly ones of relatively low, compact form, are easily thrown, balls and spheres being a delight to produce in this way.

Turning thermoplastics in their heat-softened state opens wide vistas for the development of sculptural designs having flowing lines and gentle contours, but it severely hampers one's ability to produce the sharp profiles one is accustomed to when working (in their cold state) metal, wood, alabaster, ivory, and other conventional lathe-working materials, being in this regard more like glass. In fact, one wishing to pursue diligently the lathe turning of thermoplastics would be well advised to delve into some of the literature on glass working, wherein valuable tricks may well be encountered. When sharp edges or lips are required in a turned piece, it is best to produce the basic form in the heat-softened material and then transfer this to another lathe (such as the model-maker's Unimat, for example) for conventional turning with the tools normally used there, the work itself being at room temperature rather than heat-softened.

Some of the finishing one needs on occasion to accomplish (when the electric-loop tool is inappropriate) can be done with the fume-hood lathe, using thin sanding strips or small files, but the tumbler shown in Fig. 26,a can also be helpful in this. This device, also useful in finishing pieces produced freehand rather than by machine, consists of a plastic chamber (lined with emery cloth or sand paper) and a motor-driven rotor having wide, abrasive-faced blades. The chamber is made from a medicine vial having a threaded cap, which can easily be removed or securely attached as required. A small fractional-horsepower motor drives the rotor through an O-ring and turned aluminum (or brass) pulleys, the device easily made with a lathe and other tools found in a home workshop.

The tumbler can be used as is, or it can be used in conjunction with additional free abrasive material, poured in to accelerate the finishing process. Whenever a well rounded surface having a matte (non-glossy) finish is required, the tumbler can be of assistance, particularly when hand sanding is difficult or tedious. Admittedly the tumbler is a luxury, however, rather than a necessity unless one's own sculptural preferences, like those of Henry Moore, Barbara Hepworth, and others of the "concretionary (in the geological sense) school", are for such excessive rounding of edges and surfaces as that found in nature as a consequence of protracted erosive action. The tumbler is a grand tool for simulating the natural processes in this regard, the actual mechanism involved being closely analogous to nature's own.

An additional word of warning about the hazards involved in applying the various techniques described in this chapter: When you begin working with heat-softened or molten plastics, particularly in any mouth-operated procedure (such as blow-forming or blow-molding, for example), exercise great care to avoid breathing the fumes or inhaling them while using the tools suggested for the purpose. Be careful when using an open flame, to avoid igniting the plastic or burning yourself. The electric-loop tool can give a nasty burn if it contacts the skin, so treat it with respect at all times and handle it with great care. The electric-hot-plate element used in one of the vacuum formers described here must be tended carefully to avoid starting a fire. Keep inflammable materials well away from it or from its swing-path, and make certain to turn it off when it is swung away from the forming chamber.

10

MOLDING AND CASTING

THE USE OF molds in which to cast components for miniature sculptural pieces in thermoplastics or other appropriate castable materials can greatly expand one's design horizons. Often a stock of components can be built up and kept on hand for use as the occasion demands, the final composition being achieved by gluing or welding the various elements together and, if necessary or desirable, modifying these elements with the electric-loop tool.

Open molds and hot-melt press molding

The hot-melt press molding process involves the following steps:
1. preparing a pattern of the form to be reproduced.
2. making a mold from the pattern.
3. removing the pattern from the finished mold.
4. casting the final piece from the mold .
5. extracting the casting for use in a sculptural composition.

Simple one-piece molds (i.e., open-faced ones) are easily made from suitable patterns (wax, plaster, clay, wood, metal, plastic, etc.) by pouring plaster or epoxy resin over them, centrifuging this to assure proper registraton of all details, and allowing the material to cure. If plaster molds are to be used, special casting plasters that yield molds of maximum strength (dental plaster and stone, Hydrocal, Fix-all, etc.) should be employed, and, if necessary, reinforcement in the form of gauze or other appropriate material should be added for extra strength.

Care should be taken, of course, to avoid undercuts that would prevent the extraction of the pattern from the finished mold, whether this be plastic or plaster. If undercuts are essential to the design, then the mold must be made in two or more pieces, quite a feasible solution to the problem in many cases if one wishes to make the additional investment in time.

Some molds may be made merely by pouring the plaster or epoxy resin over the pattern and allowing it to set, but centrifugation—using a simple, homemade sling centrifuge—greatly improves the chance of success in registering and reproducing all detail from the pattern. If, instead of plaster or epoxy resin, one has used modelling clay (oil-based, not ceramic clay) in

91

making the mold, centrifugation can be particularly valuable, because it means that a needle, which could easily damage the mold, is not needed to

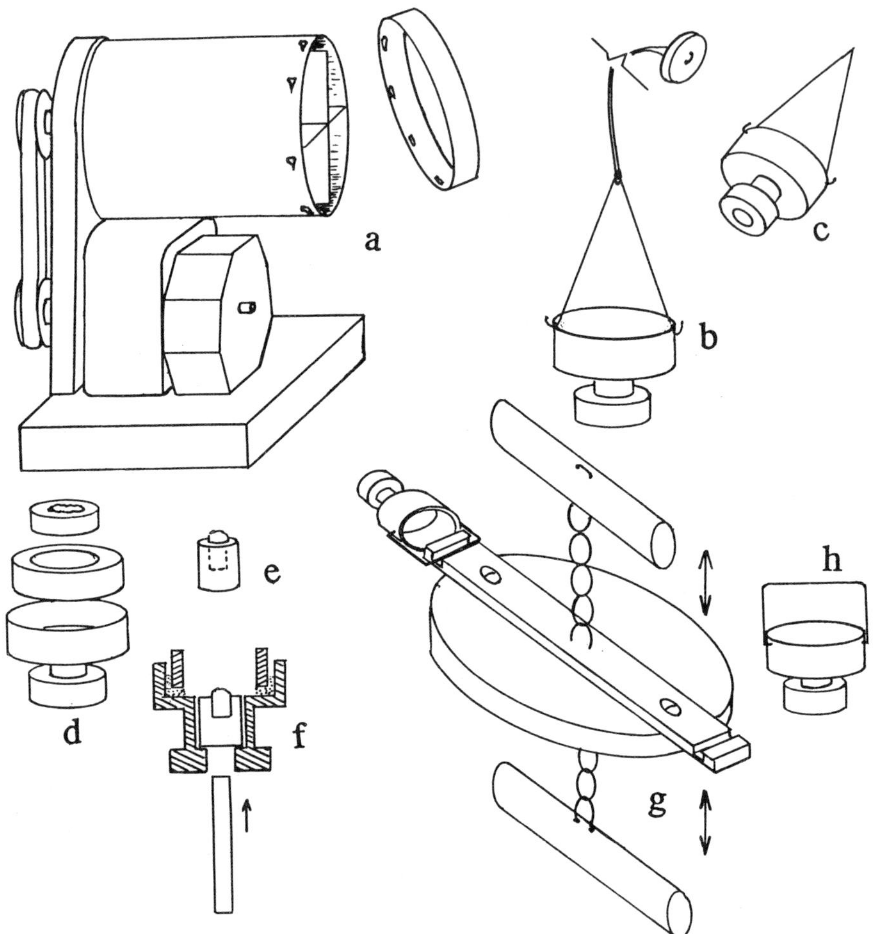

Fig. 26. Homemade molding, casting, and finishing tools for use in making components of miniature sculptural pieces. a. Miniature tumbler (5.75" high; chamber diameter: 2"; length: 2.5") for use in sanding or applying a matte finish to finished castings. b. Miniature sling centrifuge for use in mold making and in casting from molds. The body of the centrifuge consists of a liquid-dispensing cap (from a syrup bottle) fitted with a wire handle to which a string is attached for slinging. c. Centrifuge tilted to show ejection port drilled to permit one to eject (f) a mold or casting with a short length of dowelling. d. View of centrifuge body with (in ascending order) theTygon sheath used as a removable mold-producing chamber, and a mold recently cast in the assembly. e. A short dop stick and a cylinder of

remove air bubbles, the centrifuge doing this with safety and natural grace.

A simple sling centrifuge (Fig. 26, b), quite adequate to handle molds of the size generally required in making components of miniature sculptural pieces, can be made from such scrap as the dispensing cap from a syrup bottle. The one shown required only the drilling of two small holes near the upper lip (by which to attach a simple wire sling) and a larger hole through the bottom (for inserting an ejection rod when the finished mold is to be removed from the centrifuge). A length of strong twine (two feet or so) is slipped through the wire sling, and a button or turned disk of plastic is tied to the free end to facilitate slinging. Care should be exercised when slinging the centrifuge to avoid injuring some innocent bystander or oneself, the risk considerable if one's engineering or operating technique is at fault. (I minimize the risk—to others at least—by operating mine outside and in solitude.)

I find it convenient to attach the pattern-making material to a dop stick before beginning work on it, this making it easier to hold the work by hand or in a vise. When the pattern is finished, the dop stick (with pattern attached) is inserted into the centrifuge, wedged in place with modelling clay if necessary, although a shoulder left at the bottom of the cap when it was drilled for the ejection rod retains the dop stick during centrifugation. To facilitate the removal of the mold and pattern as a unit after the mold is cast in place around the pattern, I find it helpful to insert a flexible lining (in the form of a length of flexible, thick-walled plastic tubing) inside the main cavity of the centrifuge cup (Fig.26, d), securing the tubing with modelling clay as required to prevent any displacement or leakage when the mold-making material is being centrifuged. After the mold-making material has been centrifuged over and around the pattern and then allowed to harden properly, the

---

HDPE from which a miniature head or face will be carved prior to casting a mold (around and over the curved surface) in the miniature centrifuge. In *f*, the centrifuge, casting chamber, and the dop-stick-mounted piece are shown in section, and an ejection rod is ready for ejecting the dopstick, carved element, and mold [not shown; the mold cavity here shown prior to filling with the mold-forming material (epoxy cement)]. The mold-forming chamber is held in place with wax or modelling clay (dotted area), which simplifies the removal of the chamber itself from the centrifuge cup and, subsequently, the finished mold from the mold-forming cavity. g. A miniature string-propelled centrifuge designed after an old-fashioned toy long familiar to and loved by children (and adults who have retained their youthful appreciation for life's simpler pleasures). h. One of the centrifuge cups for use with *g*, merely a syrup cap (as in *b*) fitted with a bent-wire handle.

mold and pattern are removed as a unit by pressing upon them from below
with the ejection rod inserted through the hole in the bottom of the centrifuge
cup (Fig. 26, f).

When the mold has been finished, the pattern must be removed from it.
If the pattern need not be saved for subsequent use, it can simply be broken

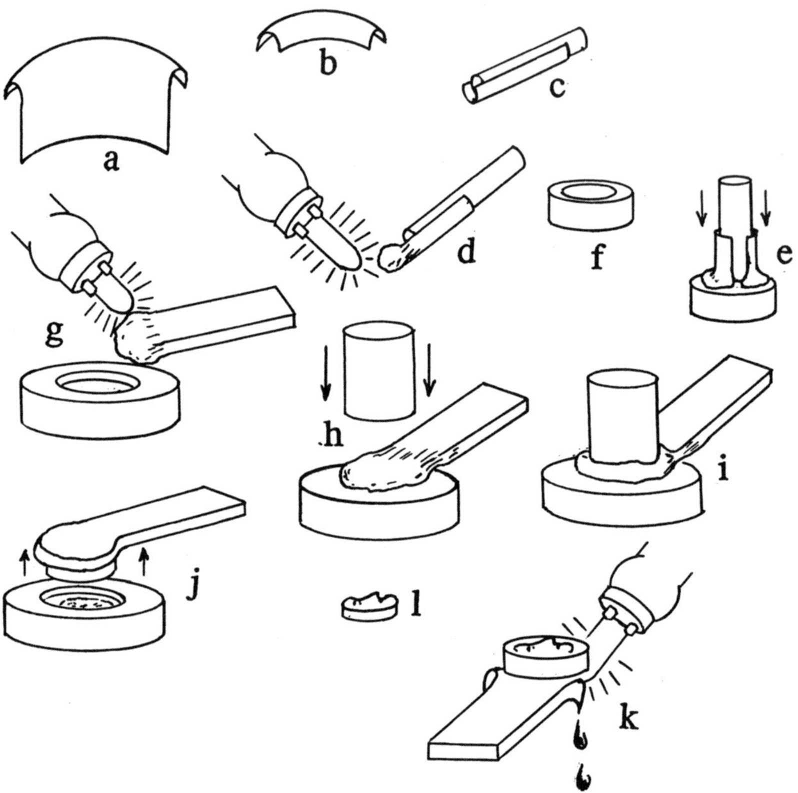

Fig. 27. Casting thermoplastic components of miniature sculptural pieces. a.
The upper rim of a yoghurt container (quart size), when trimmed away (b) and a
segment of it (c) wrapped around a length of 0.18" dowelling, makes an easily
manipulated piece of stock material with which to fill a mold, an exposed section
(d) melted with the electric-loop tool, and the dowel then used to force this molten
mass (e) into an open one-piece mold (f). A more conventional procedure is to heat
a strip of plastic stock (g) and force the molten mass into a mold (h, i) with a
tamping rod (dowelling). After the plastic has cooled, it can be flipped up and out
of the mold (j), the casting trimmed (k) from the stock piece (using the electric-
loop tool) and is then ready for use (l).

out (in the case of plaster ones) or carved out (if wood or rigid plastic), but if it must be reclaimed, an extraction rod or handle should be incorporated in the design of the pattern, and a suitable release agent (silicone spray, for example) used to simplify extraction. One advantage of using wax for the patterns (in addition to the delightful ease with which it is shaped) is that it can be melted out after the mold has cured. The mold should be cleaned well before use in casting, care being taken to avoid damaging its inner surfaces either chemically or physically.

Care should be exercised in selecting the plastic to be used for casting: one that melts easily and without charring or discoloration, as determined by preliminary tests on a small sample of the material.

The casting process (Fig. 27) involves pressing molten plastic into the mold and letting it cool there. Sufficient pressure must be applied during the process to make certain that all detail is registered in the cast piece. To assure this, I find it convenient to wrap a piece of plastic stock around a length of wooden dowelling, the segment to be melted protruding beyond the end of the dowel, the unheated section firmly gripped (by hand) against the wooden piece. (In the case of particularly small molds, a length of swab stick makes a good tamping rod.) When the plastic is melted (using the electric-loop tool), it is pressed into the mold by means of the dowel and then allowed to cool, this latter process requiring only seconds for small castings, although for larger ones it can, if desired, be accelerated by brushing the casting with cold water and then removing the water with a dry brush or small cloth. Inasmuch as the plastic one generally uses for such castings is a bit flexible even when cooled, extraction from the mold is usually a simple matter, but if problems arise, an extraction handle (a short length of stiff wire) can be fused into the exposed surface of the casting (using the electric-loop tool to melt a bit of the plastic and heat the wire before inserting it) and, after cooling, used to remove the finished casting.

Casting with matched molds

A procedure of value in producing forms not easily or quickly formed through the use of the electric-loop tool or even with simple open-faced molds involves the use of matched, or two-piece, molds to shape a sheet of stock material that has first been heated to a semi-molten state. In a sense related to yoke/plug molding, this procedure substitutes two closely matched mold halves for the yoke and plug and can achieve results more complex than those obtainable with them. Match molding, a valuable adjunct to other replication procedures, is particularly attractive in that it enables one to produce forms or elements that would otherwise require the more elaborate equipment and complex molds needed for injection molding. With this procedure, one can achieve comparable results through the use of nothing more than a clamping frame for holding the plastic blank, a small gas burner, and a pair of hard-plaster molds, the last easily made from a pattern executed in wax, clay, or various other materials. Both polyethylene (low and high density) and polypropylene are readily formed in this way, requiring relatively little heat and then cooling sufficiently slowly to give the operator time to apply and properly close the two parts of a matched mold. With a little foresight and planning, one can close the mold halves onto the plastic and apply pressure either manually or with a vise, but by standardizing the mold blanks and designing the mold halves to be closed and clamped with the aid of the drill press (used as a press and clamping device), more uniform results are easily and quickly obtained. In the simple setup shown in Fig. 28, b, a turned aluminum, acrylic, or hardwood faceplate, designed to be clamped in the drill press, has two threaded holes that permit one to secure the male half of the mold to it and be held in the chuck of the drill press while the female half is screwed to a square or rectangular block to be held by the drillpress vise. The alignment of the two mold halves for assuring proper closure and matching during the forming process is easy to achieve if the vise is clamped to the table of the press and care is taken to avoid any rotation of the chuck spindle after proper positioning has been determined by trial closures before interposing a heated plastic blank for the final molding process.

A simpler tool for holding small molds and assuring proper alignment during the molding process is a pair of pliers, preferably parallel-jawed ones as shown in Fig. 29, g. By designing the parts of the mold for attachment to the jaws of the pliers, the molding process can be made into an efficient procedure to supplement others in the sculptor's kit.

Making patterns and molds

As is true of any processes requiring the use of molds, the quality of the molds used in forming thermoplastics to be incorporated in miniature sculptural pieces depends greatly on the quality of the pattern from which the mold is made. The molds used to make such small parts as those needed here often place severe demands on the skill of the pattern maker, but with the proper tools, particularly an electric engraving tool, one can produce patterns of good quality with relative ease. The small electric tool like that shown in Fig. 28, a, when mounted in a cradle that enables one to attach it securely to the work bench at a height suitable for working with a stereoscopic magnifier (forehead mount , e.g.), makes an efficient engraving tool. A wooden spacer, carved to fit the contour of the tool, raises it to the proper height, and two adjustable hose clamps (hardward store, plumbing section), fitted to slots cut at each end of a piece of aluminum channel stock, hold it in position, the

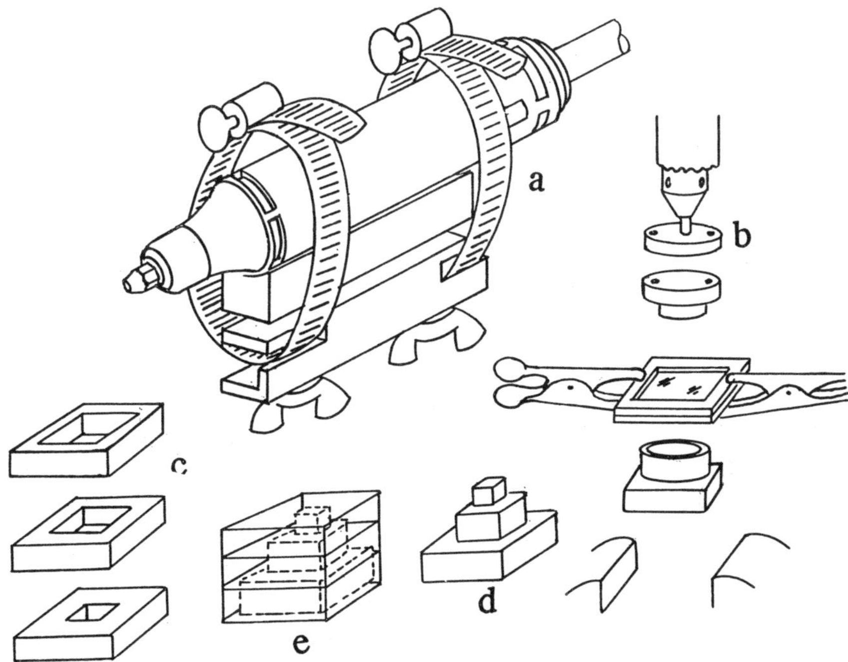

Fig. 28. Making and using patterns and molds. a. Electric tool and its mount as used in carving small patterns of elements to be incorporated in a mold. The tool is clamped (with hose clamps) in a cradle attached to the work bench, and a pattern is carved with it, using fine dental burrs. b. Molding with matched molds. The male element is attached to a plate chucked in the drill press, the female held in the jaws of a drillpress vise. The mold halves are closed with a heat-softened blank of sheet plastic sandwiched between them and thus formed by the mold, the process much like that of yoke/plug molding (Fig. 23, a). c. A laminated mold (female component), the male element (d) and the piece (dotted lines) that will be produced (e) when the mold is used (as in *b*).

cradle itself fitted with two thumb screws so that it can be secured to the work bench. Various stands and holders for such small electric tools are available commercially, but the arrangement shown here is effective, giving the operator complete freedom of movement for the use of both hands. Dental burrs make good cutting tools in making brass or aluminum patterns and molds; such burrs can be obtained in considerable variety from dental, jewelry, or craft supply stores. Small sanding disks, used with this tool, simplify the task of finishing surfaces of pieces requiring special treatment.

Rectangular, square, and cubic forms (and, by extension, buildings) are relatively easily made by pressing a suitable one-piece mold onto a blob of molten plastic to produce a solid casting. Fairly complex molds, suitable for making skyscrapers (Fig. 28, d-e, for example), may be made by gluing together a series of carved or milled acrylic laminae and using this pattern to produce a hard-plaster or epoxy mold. Various other architectural forms may be produced in the same way. Similar results may be achieved by blowing heat-softened plastic into such a mold, but in this case the finished product is merely a hollow shell, so flimsy as to be difficult to use unless relatively thick sheet stock is used initially or unless the hollow form is filled with plaster or some other material. Miniature ziggurats a fraction of an inch high are easily produced from molds, though to make the patterns for these at such a small scale can be demanding. Molds may be used in this way also to form other types of elements for sculptural compositions. The hats for the Chinese figures in Plate 11, e are easily produced by pressing a simple acrylic mold onto a molten blob of plastic, but they also can be made by blowing plastic over the mold with one of the multipurpose tools illustrated in Fig. 20, g and i.

To produce satisfactory facial features on heads measuring only a few millimeters in length can be an exceptionally difficult task, but if one, working with a suitable sculptor's wax, carves such faces and then uses these as patterns for making epoxy molds, several castings can be made from each of the small faces before the molds begin to lose significant detail (Fig. 27, j-l). Hands and feet can be made in the same way and then merely blended onto the limbs, using the electric-loop tool.

Note: Haywood (in Strong, 1938) covers nicely the essentials of molding and casting, including the lost-wax procedure. Clark (1940) treats the subject of casting and molding in considerable detail in his useful book.

11

OTHER FORMING AND REPLICATIVE PROCEDURES

SCULPTORS HAVE long analogized the basic sculptural procedures with the mathematical processes of subtraction and addition, pure sculpture (in its etymological sense) being a *subtractive* process (i.e., the removal of material from an initial mass), modelling being in essence an *additive* one. This analogy can be extended to include replicative processes, in which one element is either *multiplied* or sub*divided* to produce many.

Some of the sculptural miniatures illustrated in this book required the use of modular, or unit, construction, this involving the replication, or *multiplication*, of elements, the production of two or more identical copies of a form or shape. Less frequently, *divisional* procedures have been applied, a case in point being Plate 6, i (*The Crossbye Prime of Slakes*). A healthy appreciation for the challenge and potential of replicative techniques—in rudimentary forms probably dating back at least to the time when primitive man first came to notice and appreciate the significance of human footprints in mud, sand, snow, or dust—should be developed early and win a secure position in the artistic makeup and technical accomplishments of the would-be sculptural miniaturist with an addiction to the use of thermoplastic materi-als. One wisely turns to various fields of productive and creative endeavor—to various crafts and industries—for tools and techniques that can be applied directly to or modified for use in the production of modular sculptural elements through the basic operations of molding and casting, stamping and embossing, impressing, cutting and shearing. From the varied assortment of molds, dies, stamps, and punches in all fields of craft or industrial activity, one can select those most readily useful for the task at hand: paper and leather punches (single and turret-head styles), arch punches (and the jeweller's arch-punch subpresses), cutting punch/die units, dapping dies and blocks, hollow drills, gasket cutters, embossing, engraving and printing stamps, and dies of various types. (The potential list is long indeed, as can be seen from browsing through the appropriate catalogues of tool manufacturers). These often may be used in the conventional form or they may be modified or their improvised equivalent be made in the home workshop.

99

Punches and dies

Punches and dies of both the cutting or shearing and the shaping type
may be used to replicate parts or elements or the anlagen of these for eventual
use in a sculptural composition. Among the commercially-produced tools of
potential value in miniature thermoplastics sculpture, and in addition to the
one's mentioned in the preceeding paragraph, are the small metal punches
and shears designed for the home and craft workshop, cork borers (laboratory supply houses), chassis punches used in various metal crafts, and
washer and shim-stock punches. In my workshop, a small lever-operated
punch with a set of punches and dies for making round, square, rectangular,
and hexagonal holes in sheet metal, purchased quite inexpensively from a
tool catalogue many years ago, has been invaluable in developing some of the
elements in sculptural pieces.

In the home workshop, one can make useful punches and dies in considerable variety. Using a length of steel strapping salvaged from a packing
crate, one can produce efficient punches for working sheet-plastic stock. A
triangular cutter was made from such strapping cut and bent to the proper
shape, the cutting edges sharpened on a bench grinder before shaping. The
cutter is used in a machinist's vise, simply pressed into the plastic stock
(backed by another sheet of material of equivalent or greater hardness than
the plastic, to assure a clean cut while protecting the sharpened edges from
contact with the vise itself). Similarly, a set of hollow drills, made by
sharpening and hardening one end of a length of metal tubing in a range of
diameters, can be used to produce plastic disks suitable for various purposes,
not the least important being the blanks used with the multipurpose tools
shown in Figs. 20, g and i. Notice that the cutting edge of such hollow drills
should be notched (prior to hardening), the small notches (Fig. 20, h) aiding
in the extraction of the plastic disk with a small needle tool. Not to be
overlooked is the potential of using various cutters, hollow drills, and dies as
small molds themselves into which molten plastic can be pressed to produce
castings in a way roughly comparable to the more sophisticated procedures
of injection molding and extrusion.

Stamping and embossing

The tools for stamping and embossing may be used to produce designs
and forms either on small pieces of heat-softened flat stock or on molten

blobs of material, or, conversely, the tools themselves may be heated and applied to cold plastic pieces or masses, the choice of approach depending upon the demands of the particular job and circumstances. One relatively minor disadvantage of stamping a molten mass rather than a flat sheet is that the waste to be trimmed from the former is generally thicker and more difficult to remove. If the tools have been used in conjunction with blow forming, however, trimming is usually quite easy, because of the extreme thinness of most blown pieces.

Impressions on plastic may be made either by embossing or engraving the desired features through:

1. hand stamping, using either a single-faced stamp or a rolling one (the design replicated around the cylinder).
2. simple tool-assisted stamping (pliers, clamp, or vise to apply the necessary pressure).
3. mechanical pressing, using either a drill press or other type of press (a simple homemade one, as shown in Fig. 29, i).
4. drape stamping (heat-softened plastic either blown or sucked against the stamp, using the multipurpose tools shown in Figs.20, g and i.

The stamps used for making impressions may be such improvised or found objects as nails, screws, rivets, etc., or they may be designed and fabricated specifically to meet some particular need of the creative artist. Working at the diminutive scale of the sculptures requires some skill and finesse on the part of the stamp or die maker, not only in using the carving or engraving tools commercially available but in making appropriate ones that may be needed but are not easy to find for purchase. Dental tools, reshaped for the purpose, are most valuable, as, of course, are small tools from various crafts, but for the more demanding tasks, the electric engraving tool described earlier (see Fig. 28, a) can be a godsend, particularly when fitted with a speed-control device (I use a Powerstat variable transformer, but a foot-operated rheostat should be adequate too).

One of the simplest ways of stamping a pattern on heat-softened plastic is to use a pair of pliers having grooved jaws. The grooves may be simple paralled ones or more complex (two diagonally intersecting sets).

A simple stamping or molding press may easily be made from a clothes-pin of the type shown in Fig. 29, d. Disassembled, its jaws are modified, as shown here, by sawing and filing so that the two matching halves of a hardwood mold or stamp may be fixed in place by means of small screws. The clothespin is then reassembled and is ready for use. A small sheet of

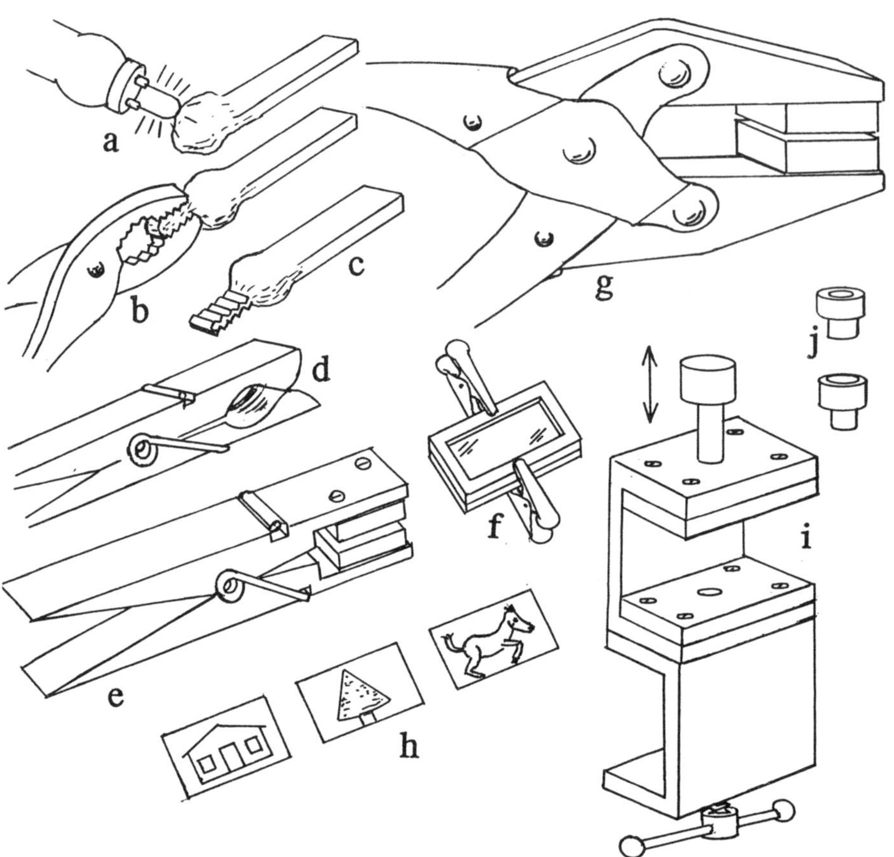

Fig. 29. Molding and stamping processes and tools. a-c. Simple procedure for shaping heat-softened plastic with pliers. Plastic stock (heated with the electric-loop tool) is clamped between the grooved jaws of a pair of pliers (b) and allowed to cool, after which the shaped section (c) is melted or cut from the stock piece and is then ready for use, a simple way of producing crenulated or corrugated elements of sculptural miniatures. d-f, i-j. Tools for matchmolding small components. A spring clothespin (d), modified as shown (e) to accept the two halves of a mold, makes a simple but effective mold-closing tool, as does a pair of parallel-jawed pliers (g), these more elegant and precise than the clothespin. (See also Fig. 30 for another mold-closing device.) The frame for holding sheet stock (f) fits between the clothespin jaws without hampering proper closure upon the heat-softened plastic sheet in producing such formed pieces as the ones shown in *h* (each only 0.5"x 0.37"). i. A simple, easily-made press suitable for closing matched molds (like the two halves shown in j) and in other operations, e.g., stamping.

plastic stock, clamped (alligator clips) between two brass frames, is heat-softened over the flame of a miniature Bunsen burner and then positioned between the open jaws so that the mold halves can be closed onto it. Note that in modifying the clothespin jaws sufficient space must be left behind the mold halves to accommodate the stock-clamping frames when the jaws are closed onto the heat-softened plastic.

The stamps or molds used with this device may be made by sawing, filing, or carving, but an effective way of producing a wide range of forms is to build the male form up with quick-setting epoxy cement applied over a pencilled sketch directly on the face of the mold block, transfer the design to the matching block, and incise the female form with burrs held in a hand-held or bench-mounted (Fig. 28, a) electric tool. The two mold halves are then screwed to the jaws of the clothespin, and the device is ready for use. The range of forms reproducible in this way is vast indeed, and one's skill at producing molds increases rapidly with practice, allowing one to broaden considerably the scope of one's design horizons in most challenging and fascinating ways—all at negligible expense!

In selecting a clothespin for the job, choose one with good jaw alignment, so that the two mold halves close securely upon one another with accurate registration across their faces. One or two alignment pins may be added to achieve this. A sturdier, all-metal clamp and one with excellent alignment , is the Stanley no. 43-161P, obtainable at hardware stores. The two boxwood mold halves are secured by screws to the jaws of the clamp (Fig. 30, a), ample space left behind the mold blocks to accommodate the rear section of the blank-holding frame during the molding operation (Fig. 30, b).

Simple flat stamps or dies—as opposed to ones designed for rolling—may be applied through the use of a drill press, a small homemade press like the one shown in Fig. 29, i, or with a pair of pliers (preferably one with parallel jaws (Fig. 29, g), the stamping or die elements designed to be held securely by whichever of these is selected. The small press, initially intended for use in thermoforming sheet stock, consists of two short lengths of aluminum channel stock screwed together and fitted with a screw and handle for attaching it to the work bench. Alternatively it could be designed to be held by a vise if preferred, the essential requirement being stability. Two metal plates give thickness to the jaws of the press, the lower plate drilled to accept a holder for the female element of a two-piece stamp or die, the upper jaw assembly drilled for the plunger to which the male component is attached. The size of the miniature press makes it convenient for use close to the work area, where space might be inadequate for a drill press or other large press tool. Another mold-closing device, this based on a metal clamp of the type readily available in hardware stores and easily modified for the task, is shown in Fig. 30.

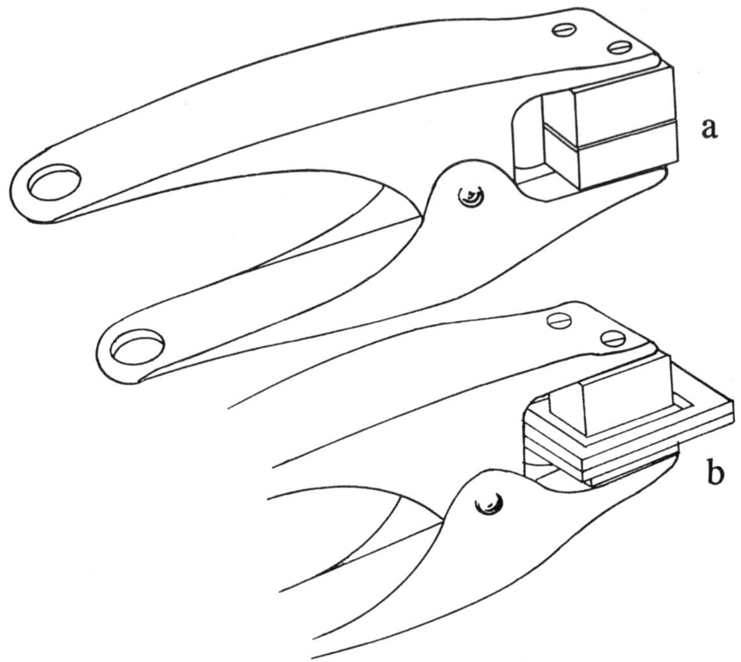

Fig. 30. Clamp for closing matched molds. a. Wood-working clamp with two mold halves attached (by screws in holes drilled for the purpose) to its jaws. b. Clamp in use, the two mold halves here shown closed on a heat-softened plastic blank held in a frame like that shown in Fig. 29, f (two alligator clamps shown there omitted from the drawing in the interest of clarity).

Rolling stamps, made by engraving the desired features around the circumference of a rod of metal, hardwood (boxwood, e.g.) or plastic, can be a useful addition to one's kit of replicative tools. A method for making miniature hands having separated or outstretched fingers, as an alternative to the one described earlier and requiring an armature (see Fig.11, f), involves the use of a rolling stamp made from the tapered cap of a toothpaste tube, a wooden handle glued in place (Fig. 31, g) by epoxy cement. The stamp is rolled into and through some urschleim (unformed blob of molten plastic) to produce a corrugated ribbon from which the individual rodlets can then be cut and applied to a small palmate piece of the same material, or, more simply, the roller can be applied in such a way (Fig.31, e) as to leave an intact margin that serves as an anlage from which to shape (with the electric-loop tool) the palm of the finished hand. The rolling stamp used in this operation was made by gluing the cap of an ointment tube to a short length of dowelling, the critical feature being that the cap itself be cylindrical rather than conical or tapered. A good collection of caps of various types, saved from tubes and bottles to be discarded, can be a valuable source of ready-made stamps and dies, or they can be modified in various ways to meet the sculptor's needs. If suitable caps are not available, one can, of course, make one's own rolling stamps by sawing, filing, or otherwise engraving and embossing metal, wooden, or plastic rods. By using a metal mandrel, heated for the purpose, one can form miniature pleated skirts or other garments with folds, but the process requires a bit of practice before one can expect acceptable results.

One of the advantages of thermoplastics over stone as a sculptural material, of course, is that it is possible with them easily to achieve the separation of small or delicate elements from a larger mass or from each other (as the outstretched fingers on a hand, or an arm raised above the head). Such separation is simple to achieve through the use of stamps and dies of various types, both flat and rolling types, the amount of separation obtained depending upon the amount of pressure applied during the stamping or pressing operation.

A special use of rolling tools is in producing flat stock in thicknesses not readily available, stock thinner, for example, than that normally encountered in food containers. By rolling a suitable tool (metal rod) over a molten blob of plastic positioned between two thickness-regulating strips of wood, metal or plastic, the desired thickness can be obtained. Alternatively, one can use a jeweller's rolling mill, but these are expensive. The homemade mill shown in Fig. 31, i will do the job, costs little to make (if one has the machine tools needed), and will give much satisfaction, both in its construction and its use. Made originally for flattening wire stock for use in making the hardware for model-ship gunports and doors, this tool enables one quickly to build up a stock of flat strips in dimensions appropriate to miniature sculptural pieces. The tensile strength of polyethylene and polypropylene is so much less than that of brass or iron wire, however, that great care must be taken to avoid breaking the material as it is pulled through the rollers; skill comes with practice. To use the tool, a length of stock is fed through a hole at the rear until it passes between the rollers and can be grasped with a pair of flat-jawed pliers. The rollers are then tightened gently and the plastic pulled through, being flattened slightly in the process. The procedure is then repeated as often as required to produce the desired thickness. One soon learns

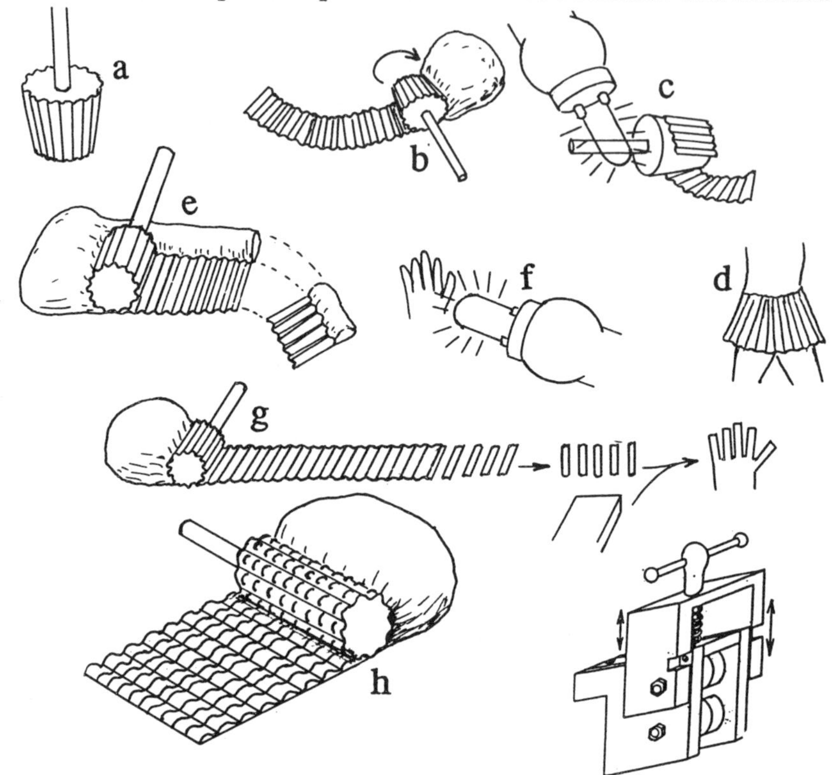

Fig. 31. Rolling stamps and the stamping process. a. Rolling stamp, made by gluing (epoxy cement) a length of 0.125" dowelling inside a toothpaste-tube cap. b. The stamp in use, rolling out a corrugated strip from a heat-softened mass of polyethylene. c. Using the electric-loop tool to heat a brass mandrel around which a corrugated strip of plastic (produced as in *b*) will be shaped for fitting onto a miniature figure (d) in a sculptural composition. e. Rolling a strip of plastic to form the anlage of a miniature hand. Notice that the rolling stamp in this case was made to leave an unstamped section of plastic which will later become the palm and wrist of the hand. f. Fusing and shaping the palm and fingers to make the finished hand, using the electric-loop tool. g. An alternative way of producing a hand, by rolling out the fingers, separating some from the rolled strip, and building up the palm with fresh stock. h. Stamping a section of simulated tile roofing through the use of a special roller machined for the purpose in the home workshop. i. Homemade rolling mill for flattening wire or round plastic stock.

the feel of properly applied pressure as one draws the material through the rollers. Needless to say, the tool works well (and requires less care) with wire, when flattened metal strips are required in one's sculptural compositions.

Cameo production

An efficient and effective way of making sculptural pieces in cameo style (bas relief) is to build the plastic up directly onto a picture or drawing of the image to be reproduced. This illustration of the contemplated work should be the actual size of the finished piece, as the plastic will be melted into place over its surface to capture all the detail. The illustration should be mounted on a rigid piece of cardboard (posterboard, for example) to provide adequate support for the molten plastic as it is made to flow into position atop the illustration. Because the plastic in thin sections is brittle and fragile, it is wise to embed segments of fine wire in such areas to prevent breakage when the finished piece is removed from its backing and prepared for final mounting and display.

The paper surface onto which the plastic will be melted should be given a coat of vaseline or other type of release agent that will prevent the plastic from flowing into the grain of the paper. If the surface is properly treated, the eventual separation of the sculptural piece from its temporary backing will be a simple matter. A sharp, thin-bladed tool (fine scalpel, razor blade, craft knife, etc.) should be used to effect separation, particular care being exercised in the thiner and more delicate areas of the work.

To maintain and take best advantage of the cameo effect when displaying and viewing the sculptural piece, it should be mounted on a piece of posterboard of contrasting color, blue being excellent, for example, with pieces made of the white or ivory-colored polyethylene used in many yoghurt and margarine containers. In this way, the effect achieved in classical cameos (through the skillful use of the color-layering found in natural stones) may be simulated quite strikingly. By mounting the piece on short pedestals of hot-melt glue and polyethylene that raise it a millimeter or so above the surface of the posterboard, the illusion of depth is increased and the overall effect further enhanced.

One interested in experimentation might well be intrigued by attempting color-layering or lamination in the thermoplastics themselves to simulate that found in natural stone. A simple way of producing a block in which two or

more layers of plastic stock in contrasting colors are laminated into a unit is
to clamp strips of plastic together between metal plates and heat the sandwich until the surfaces of the layers fuse slightly. When allowed to cool and
removed from the clamping plates, the block can then be "carved" (with the
electric-loop tool) to produce raised or incised designs in a color contrasting
with that of the layer in which the design appears.

Painting with thermoplastics

When one has become adept at producing the cameo effect just considered, the step to painting with thermoplastics is a simple and natural one. At
this stage in one's technical development, the close relationship between
sculpture and painting becomes most apparent, and the challenge of branch-
ing out from miniature sculpture to the painting of miniatures may well prove
irresistible. My only advice at this point is that you attempt a few simple
paintings in order to develop a feel for the procedures involved and that you
then let your own impulses and preferences direct your course.

Platt (1985), in his study of Medieval Man (pp. 156-7), pointed out the
importance of the interplay between painting and sculpture in the development of fourteenth-century art. By the time one reaches the middle ages in the
development of one's own technical skill as a thermoplastics miniaturist, one
is ready to attempt the fusion of painting with sculpture, a step having been
made in this direction with the built-up overlay process just described in the
section on cameo production. A delightful exercise along such lines is to
attempt to copy some of the simpler "creatures" and objects Kandinsky used
in his charming *Sky Blue*. Begin by attempting to reproduce the forms in
opaque white HDPE, ignoring the colors until you are satisfied with the
forms themselves. When your forms look like his, try reproducing them in
colors approximating his, resorting ultimately to acrylics (painted onto the
plastic pieces) if you are unable to handle the more demanding task of mixing
colored thermoplastics. Once you master the simple forms, have a go at some
of the more complex ones, but be prepared for disappointment along the way.
It is not an easy task in these materials! You will, however, probably find the
exercise so challenging as to be led off into creative directions of you own. In
any case you will be a better technician than you were before the exercise
began.

*Remember to exercise great care when heating plastics. The dangers
of fire and burns and the hazards of toxic fumes are real!*

PART FOUR

IN PURSUIT OF NUMEN

(SEE ALSO: MUSICAL APPENDIX)

THE LEARNER

As we try to learn the sculptor's skills
By copying the work of the sculptural great,
We, too, experience creative thrills
That all in art appreciate.

To make a copy in miniature
Of the *Nike of Samothrace,*
(A piece like it that will endure,)
As a study is no disgrace.

By copying our predecessors' work,
We learn the tricks of the trade;
By learning never toil to shirk,
We find in truth how art is made.

At first one's copies are made by rote,
Each stroke of the master copied with care,
But once one developes a skill of note,
Something creative we are ready to dare.

And dare one does to try something new,
Something that differs from all the rest;
Ready at last to strive with the few
Who are only content with the best.

So one learns from those student days
A skill that leads to the heights
And helps one acquire the artistic ways
Of producing creative delights.

GENERALITIES

MUCH HAS been said of the goals and aims of art in general and of sculpture in particular, of those attributes that should be considered in creating a sculptural composition. Thus, balance, unity, diversity, contrast, universal harmony, and truth to the materials, all must find a place in setting goals of design and execution; but the history of this art—as of art generally—clearly indicates that there is, indeed, a time and a place for all things: temperance, intemperance, restraint and lack of it, simplicity, complexity, movement, inactivity, imitation and originality, simple flights of fancy and complex ones.

To paraphrase Sir Joshua Reynolds in his tenth discourse on art, directed specifically to sculpture, color, for example, should be applied to sculpture only if the aim is, as he says " to administer pleasure to ignorance or be a mere entertainment to the senses", because sculpture has a higher aim and provides a nobler delight, the intellectual one of contemplating the perfect beauty of an elegant form, such elegance being one of sculpture's highest excellencies. The purist who agrees with Reynolds that elegance of form takes precedence over all else in sculpture—although, as Cook (1972, p. 75) reminded us, color was extensively used on the sculptural pieces of classic Greek where white marble remained the material of choice, the supreme material—will prefer to keep the miniature sculptural pieces in opaque white polyethylene or polypropylene, materials closely resembling (in some important aspects at least) the white marble most admired in Greek pieces even today. It seems, however, that part of the mystique of the modern thermoplastics—deriving in no small measure from their potential for plastic flow when molten—is the natural and convincing way in which they can be carved and modelled, and, through the use of heat, made to simulate forms and features difficult or even impossible to reproduce in stone, deceptions that are entirely and easily within the more versatile range of thermoplastic sculpture and which in no way abridge or destroy such higher sculptural qualities as the excellence and sheer beauty of form. To this greater range of imitative or representational possibilities in the *form* thermoplastic sculpture can exploit should, then, be added the enhanced scope available through the use of *color*, bringing the miniature sculptures in this respect closer to painting than classical sculpture could have come if it relied only on the color inherent in the more restricted range of sculptural stone.

Miniature thermoplastics sculpture seems most magnificently and happily to refute the frequently-expressed sentiment (see Licht, p. 13, for example) that sculpture—being of monumental scope and thus capable of meeting a community need for objects of monstrous dimensions— is essentially *communal* in nature. To underscore an advocacy of the *privatization* of sculpture, this book emphasizes the way in which one can through it satisfy one's own need for a private, personal, and individual expression of the artistic urge present in so many of us but logistically more difficult to develop with conventional tools, materials, and dimensions. It is primarily directed toward the reader with autarchic tendences, one craving a high degree of self reliance and independence, not requiring a team of helpers and associates, but quite able, with scrap material and simple homemade tools, to achieve worthwhile goals without external assistance. Fortunately, however, in the early stages of acquiring and developing the necessary skills for some of the more sophisticated technical procedures, one must develop an eclecticism and dependence that quickly dissipate any natural tendency toward solipsist arrogance.

To strive for moderation in all things and at all times in one's efforts is to stultify the potential the art has for satisfying one's own technological cravings, for stimulating others who view it, and for soaring to heights of creativity initially undreamt of, so, in the interest of depicting something of the potential for technological expression in miniature thermoplastics sculpture, consider for a moment the character of the numina of this particular art form.

The numina of thermoplastic materials in general and in their particular application to miniature thermoplastic sculpture, those qualities—mysterious in toto—that appeal to the aesthetic sensitivities and that, in the hands of a master, are, on occasion, even able to hint at the existence or presence of a spiritual or even a quasi-divine aura, include the following:

1. A vast potential for highly diversified morphological expression.
2. Horizon-broadening and eclecticism-stimulating versatility.
3. Benevolent technical logistics (mastery relatively easy to achieve).
4. A special environmental friendliness or verdant promise, stemming both from miniaturization and from the nature of the materials themselves.
5. A hopeful, promising outlook for future development, deriving in part from the nature of the materials but in part the legacy of the history of the plastics success story.
6. The mystique of miniatures.
7. The tractable, resilient, and forgiving nature of the materials.

12

THE TECHNICAL MEANS
FOR ACHIEVING DIVERSITY

THE PRINCIPAL CHARACTERISTICS subject to variation in creating miniature thermoplastics sculpture are the chemical composition and physical behavior of the plastic materials, the texture (both surficial and internal), the form or shape of the component parts and of the whole composition, and the color of these last.

The chemical composition of the plastics themselves is set by the manufacturer and may not require alteration by the sculptor, but the potential for such alteration exists and the advantage of doing so can, in some cases, be apparent as, for example, when one adds wax- or oil-based pigments, hot-melt glue, or mixes one plastic with another in the molten state. Generally speaking, however, chemical alteration is not the primary goal but, rather, an inevitable consequence of the adulteration one attempts.

By contrast, one often sets out to modify the physical properties or behavioral characteristics set by these properties, this in an effort to improve the working properties in the pursuit of one's sculptural goals. Thus, the ductility of the plastic can be altered in an intentional and purposeful way, for example, by adding hot-melt glue if before the addition a piece refused to be drawn out as easily or as far (without breaking) as one would like.

The surface finish of the plastics recommended earlier as stocks for the miniature sculpture tends to be smooth and even slippery, often with a lubricity of its own. In most cases this is not objectionable; indeed, at times it may be highly desirable, as, for example, when the smooth metallic surface of a mold, stamp, or die must be reproduced on the heat-softened plastic these are to impress. Most other types of surface finish, however, must be imposed by the sculptor. Mention has already been made of the use of a miniature tumbler for use in giving a sanded, or matte, finish to elements that will be incorporated in sculptural pieces. Similarly, sawdust, sand, powdered minerals of various types, and other pulverized or powdered materials can be added to the plastic in its heat-softened stage to modify the smooth finish; with suitably finished molds, stamps and dies one also can accomplish this. Elaborate tools are not necessary to achieve some effects that can be quite useful, however. A simple way of producing a hummocky surface, for

113

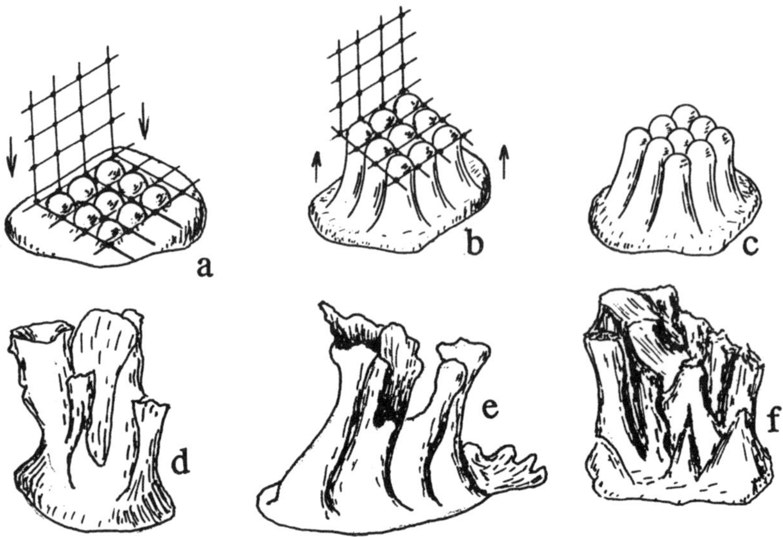

Fig. 32. Using hardware cloth to produce anlagen for a miniature sculptural piece, an example of *divisional (subdivisional)* sculpture. a. A piece of hardward cloth (0.25" meshes) is pressed into a molten mass of HDPE and then (b) withdrawn upward to produce a hummocky anlagal mass (c) which can be developed in various ways (d-f), through the use of the electric-loop tool, to create sculptural pieces.

example, is to press a strip of hardware cloth onto the heat-softened plastic until a small hummock develops in each opening in the metal strip (avoid overheating to the point that the individual hummocks flow into one another). When the mass is cool, the hardware cloth can be lifted off, leaving the hummocks as distinctive and conspicuous surface modifications (Fig. 32, and Plate 6, i (*The Crossbye Prime of Slakes*). In a similar way, a rough granular texture may be imparted to an otherwise glabrous surface by pressing a piece of sandpaper (its grit size appropriate to the scale of the piece being worked) against the heat-softened surface and allowing it to cool in place before removing.

In a similar way, various other objects can be used as stamps to impress designs on the surface of the plastic. Thus, for example, a wood rasp, when pressed into a heat-softened surface, imparts a scaley texture, and a file a series of closely-spaced and uniform lines, capable of replication *ad infinitum*. Such inexpensive commercially available objects as wood or machine screws and rivets or paperclips make interesting tools for producing unusual designs on the surface of heat-softened plastic as one developes sculptural pieces.

Special sintering and spluttering equipment may be interesting to investigate, but a less expensive way of producing their effects is to heat plastic (in a metal container) and pour it onto a stiff piece of paper or a piece of wood, preferably open-grained, such as Philippine mahogany. If the plastic is sufficiently hot when poured, intense bubbling results; if the molten mass is then sandwiched and pulled (as described for the pulled-adhesion procedure (p. 57), interesting anastomosing patterns develope around the attached margins of the plastic and can be used to simulate fimbriate organic structures. More effective means for achieving these and other potentially useful textural modifications could, in all likelihood, be developed after some experimentation, but *in all of this one should be aware of the hazards inherent in working with hot plastics and take every possible safety precaution.*

Numerous ways of modifying the form, or shape, of plastic materials have been treated in earlier chapters, but numerous additional procedures, such as the use of the lathe's knurling tool (to make fluting), and of embossing rollers in a lathe (rather than hand rolled), could also well be explored.

The aesthetic appeal of opaque white sculptural pieces mounted on a blue background of high saturation and brilliance is considerable, such pieces having much of the charm of carved ivory, and the range of subjects which can be depicted in such elegant style is inexhaustibly vast, but those craving too the entrancing charm of color in the actual sculptural piece itself will find it easily achieved when working with thermoplastics. The vast potential for using and controlling color and for partaking of its benefits is in one important sense a unique feature of the numen of thermoplastics. A feature not found in stone, wood, or ivory, but present in wax and clay is that color can be added not only *surficially* but, when the material is in its molten state, *throughout* the mass, to become an integral part of *all* the material rather than a mere veneer.

In contemplating the genesis of variegated sculptural masses wrought through the heat of the electric-loop tool, it is tempting —and with considerable justification, inasmuch as such phenomena as plastic flow, convection currents, and mineral segregation are involved in both—to compare the processes and results seen in the genesis of various marbles and marble-like rocks classically favored by sculptors down through the ages. Nature has done this with all of the conventional sculptural materials, but it is difficult for the sculptor to simulate it in them if it is not already present. With the thermoplastics, however, it is a simple matter to introduce color in depth, because in their molten state the plastics permit one to swirl color in and distribute it throughout the mass as desired. Nature does this mixing in stone, wood, and ivory as these materials develope, whether as molten masses, in the inanimate growth of crystals, or in the organic growth and deterioration (as in the spalting of wood, for example) of plants (wood) and animals (ivory).

In analogy with natural processes, a highly effective way, for example, of obtaining the streaks and splotches of color simulating those in marble is to use scrap plastic stock that has words or colored areas printed on it. As the stock is melted, the contrasting color from the printed areas can be blended selectively into the sculptural piece being produced. Thus, in attempting, as a learning experience, to make a miniature copy of Étienne Hajdu's *Delphine*, it was possible to add swatches of color to match those on his marble sculpture in this way, although it could also have been done by using powdered graphite (pencil scrapings). The scope for introducing diversity along such lines is broad indeed for one prepared to undertake some experimentation in the choice of colorants and adulterants as well as in the tools and techniques for achieving suitable mixing patterns. One may accumulate stocks of colored polyethylene to use directly in the work, one may blend (including by coextrusion) to varying degrees plastics of different color to produce anything from incomplete swirling mixtures of the two or a new color entirely, or one may add color from separate sources to prepare one's own working stocks in the colors one wishes, the principal consideration being that such adulteration does not affect the working properties adversely.

The effect of blending plastics of different color can be determined in advance by testing such properties as ductility in a small sample with the electric-loop tool and needle. One should, of course, avoid plastics that tend to char or that are brittle. It sometimes is simpler to add one's own pigment to white stock, preferably at the time of extruding rods, threads or strings for subsequent use. An inexpensive and satisfactory source of such pigments is the wax crayon, although artist's oil-based paints and acrylics should also be considered (first tested for compatibility). The electric-loop tool can be used quite effectively to blend color into plastic stock when working with small quantities of material, but if the volume is too large for this and a gas burner must be used, the wax crayon can simply be brushed onto the warmed plastic mass before melting occurs and then mixed into it as soon as the mass melts, the degree of mixing depending upon the results one desires in the finished blend.

13

BROADENING DESIGN HORIZONS AND ENCOURAGING ECLECTICISM

The sources of sculptural ideas

WE, IN OUR EVERYDAY LIVES, are surrounded by forms that can be adopted directly or with modifications for use in making our miniature sculptural compositions. The creation of representational pieces based on what we actually see could alone occupy the sculptor's time, the possibilities endless. A few of the many categories of commonplace things and places suitable for transformation into miniature sculptural pieces and from some of which examples have been included in the plates in Part Six are:

dancers, skaters	workmen at their work
acrobats	artists, sculptors
animals	plants
windmills, bridges	lighthouses
musicians	actors
circus scenes	domestic scenes
characters from literature	political caricatures
cartoon characters	jewelry, cameos, medals
portraits (busts)	landscapes

Is it any more risky for us to say "There's art in all things, if man had eyes" than it was for Byron to have said "There's music in all things, if man had ears"? Certainly not. Not only may art be seen in the actual images that meet our eye, it may also be seen, as artists have known down through history, in the form of latent images, not seen in the ordinary way but in different ways, as adaptations leavened by imagination from the merest suggestion of an alternative image. And so, from the boundless realms of imagination and fantasy—even fantasy unprompted by actual images and the creation of the imagination alone (is it ever alone, devoid of previous influences of some sort?)—can come (to distort Shakespeare for our purposes) more images from heaven and earth than are dreamt of in our many and diverse philosophies about art.

117

Just as Leonardo was able to see battles in cloud formations and he and Pietro di Lorenzo were able to derive images for their own use from gazing intently at a smear of spittle or the markings left by plant growth on a stone wall, so, too, can we find pictures in the decaying venation of a dead leaf, in the jagged edges of a tissue-thin slice of French bread, in the folds of a valance, or in a haphazard coil of wire or rope. We can take our cue to the alternative interpretation of commonplace images from Mauritz Escher, who derived marvelous stimulation from crystal forms and structure and, when his imagination flagged, from listening to Bach's music, which revitalized it and set it off on renewed flights of fancy. We can, like Max Ernst, derive creative images from *frottage*, make our own rubbings and use them to stimulate our imagination.

One of the delights of myopic vision (and one of the sweet uses of adversity) is the ability to create a new world of images—fantastic delights (faces in flowers, figures in leaves and bushes)— merely by taking off one's glasses, to produce a world perhaps a bit blurred but one filled with images half seen or imperfectly seen yet there and ready for adoption and adaptation for one's sculptural needs.

I happen to have spent my professional life looking through a microscope and thereby inserting myself into the company of one of the oceans' most beautiful assortments of lilliputian shell-producing organisms, the Foraminfera, a highly diverse array of forms of potential use to the sculptor seeking ideas from a relatively unexplored realm. Some of these forms, derived from enlarged models I made years ago as teaching aids, have crept into certain of the pieces illustrated in the plates of Part Six. The famous German biologist Ernst Haeckel was keenly aware of the beauty and artistic merit of many of the world's lesser known creatures (including Foraminifera; see, for example, his plate 81) when he assembled into 100 plates and a book (1904, Dover reprint: 1974) the magnificent drawings he had made of many of these as well as of more frequently encountered plants and animals. Looking at Jean Arp's *Star* and even his *Ptolemy I*, one wonders if he had not somewhere encountered the skeletal remains of some microscopic silicoflagellates. Might not Maria Martins also have been led by a similar encounter to create her *Rituel du Rythme*?

But we needn't have the luxury of a microscope to create useful images on which to base our sculptural efforts. A bit of scrap drawing paper can be chewed into delightful forms; a handful of popcorn often contains some charming miniature poodles, ideal for converting into popcorn poodle pets for the amusement of children (and grownups alike); a disk of peppermint (Starlight mints are good for this) develops intriguingly delicate meshworks as it is sucked into delightful oblivion; shadows cast on the refrigerator or stove by rays of sun intercepted by plants and knicknacks on a window shelf give a most marvelous puppet show (particularly when the wind is blowing), with scenes and characters just begging for conversion into sculptural miniatures; the scum formed by powdered milk in a cup of hot water, when pricked with a small knife or toothpick, can be modified to produce forms interesting for translation into sculptural pieces, particularly ones of a protean or fantastic nature; some of the pieces produced through the pulled-adhesion technique described earlier resemble the motley and slimy collection of intriguing marine organisms (particularly tunicates and ascidians) one encounters in the intertidal zone of rocky seacoasts; the shavings produced by various machine and hand tools as one works metal, plastic, wood, leather-hard ceramic clay, wax, paper, and other material often can become the basis of or suggest the form for some element to be used sculpturally; and in the shavings from a hole saw, produced as one cuts out blanks of HDPE, parts of trees and bushes appear, as if by magic, requiring only a little attention from the electric-loop tool to clarify or help develope some critical line or surface.

The English watercolorist Alexander Cozens long ago (1786) discovered the joy of using ink blots in clever ways to create images which he could developed into landscape paintings, far in advance of the psychoanalytical uses to which these were put by Rorschach and Freud or the equally or even more curious *paranoic-critical* method Dalí employed in some of his artistic creations, but the basic procedure, with some modern materials, can be of great value to the sculptural miniaturist.

Consider first, for example, an experimental procedure for converting a type of painted image to a three-dimensional, or sculptural, one. The outgrowth of the simple fortuitous observation that, like the images produced in a tea cup (and subject to delightfully imaginative interpretation by a skilled tea-leaf reader), the residue left in a translucent polypropylene cup by one's breakfast mixture of yoghurt and dark-red stewed fruit (plums, blackberries, etc.) can be highly suggestive of forms that are eminently suitable for conversion to sculptural compositions. The secret of revealing the images with clarity lies merely in positioning the cup in such a way as to take full advantage of the light applied in proper intensity and at the proper angle of incidence to set the image off in clear detail when viewed against a background of suitable contrast. If the field of images left fortuitously is not satisfactory, one can modify these readily by brushing them into more usable shapes with a spoon, toothpick, finger, pencil, needle, or other handy tool.

The images may be photographed if the facilities for this are at hand, but they may also be transferred to paper by cutting the cup into strips and xeroxing them, as was done for those in Figs. 33 and 34.

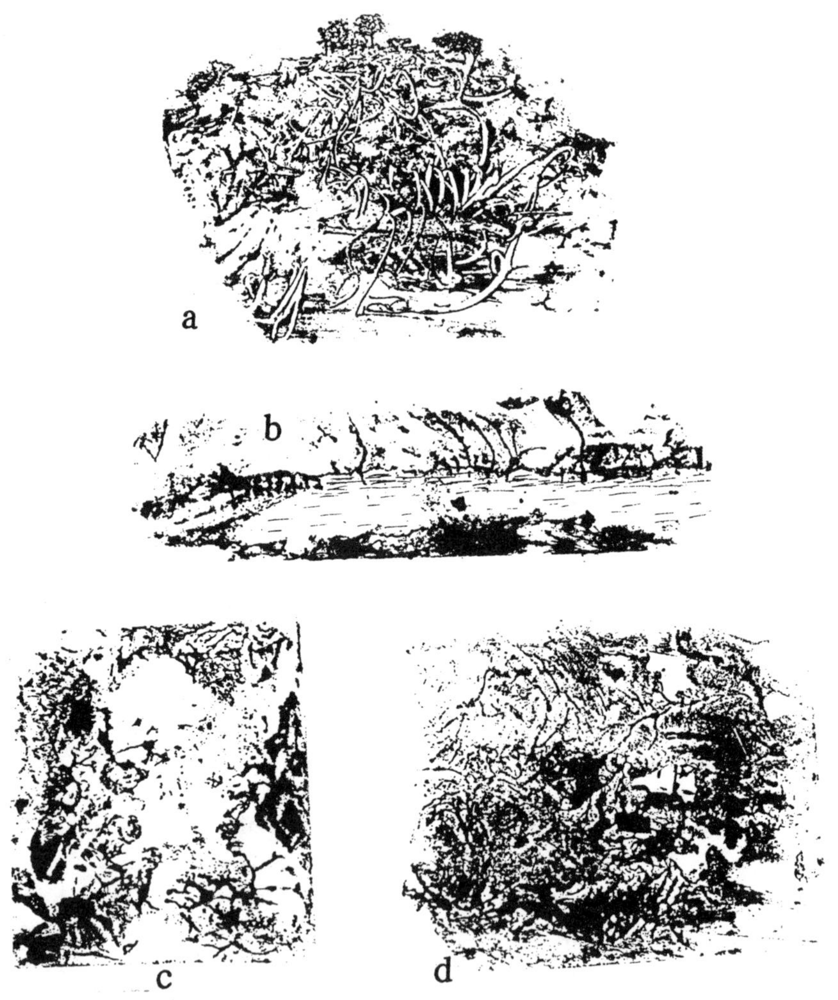

Fig. 33. Images (a-d) produced by randomly smearing yoghurt and blackberry juice around the inside wall of a transparent plastic (polypropylene) cup, cutting the cup apart and photocopying the segments thus produced.

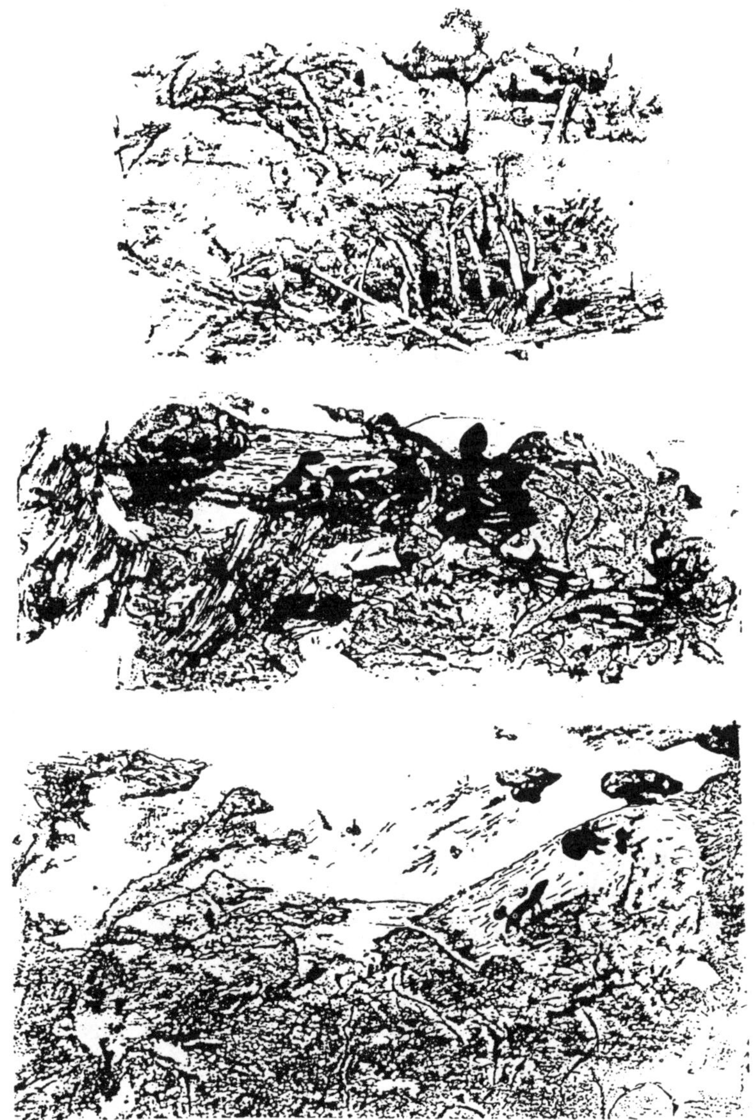

Fig. 34. More scenes produced in the same way as those shown in Fig. 33.

More importantly, though, the desire to capture the images (and improve their quality) by other means than photography or xerography led to experimentation along another line, i.e., painting images directly on clear sheet plastic with such clarity and brilliance that they could readily be used as miniature maquettes suitable for translation into sculptural pieces. A set of these images is shown in Fig. 35. The first material selected for forming such images proved to be most satisfactory, not only because they were beautifully clear, but because they were sufficiently thick to give an impression of depth as well, a useful feature in the three-dimensional interpretation that was subsequently to be attempted. Carpenter's glue (aliphatic resin), to which enough India ink was added to produce a black mixture, made the "paint" for the miniature images, and, though any one of various small tools is useful (from toothpicks and match stems to a fine-tipped brush), I use a miniature plastic spatula (Fig. 5, k) having a wooden handle (a short length of wooden stick of the type used for cotton swabs) and a 0.25-inch-wide blade (low density polyethylene), a particularly efficient tool, not only for transferring the mixture to the film but also for shaping the images that are to be produced. The mixture can be diluted with water, if desired, but in its undiluted state it gives desirable depth to the image and, of equal or even greater importance, its stringy consistency (when the mixture is drawn out or moved about on the film by means of the edge or tip of the spatula) is strikingly similar to the stringiness that characterizes properly compounded polyethylene or polypropylene when these are melted and being worked into anlagen for miniature sculptural pieces. This resemblance in the behavioral characteristics of the glue mixture and the sculptural materials creates ideal circumstances for transforming the painted image into a true (i.e., three-dimensional) sculptural form. Moreover, if one wishes to enhance the painted images by adding tonal variation and gradation, they can be made into artistic creations of some merit in themselves. This can be done by adding a small quantity of water, stirring it only lightly to leave the mixture heterogeneous. Fascinating tonal variety is automatically achieved when it is painted onto the plastic film.

Once acceptable images have been produced and allowed to dry on the plastic sheet, they may either be used directly to produce xerographic copies or photographic negatives and either contact or enlarged prints that stand as finished works on their own, or they can be used as guides in producing the sculptural form, much as maquettes are in conventional sculpture, their principal virtue in this regard being that it is so easy through them to produce a widely diverse assortment of images of use in stimulating one's own creative flights of fancy.

The images shown in Fig. 35 have been converted into the sculptural pieces with which they are paired in Plates 1 through 5. Some of them would make attractive cameos, easily and effectively done with the procedure described on pages 107-108.

Another group of fortuitously created images is shown in Fig. 36, essentially doodles, unplanned and incomplete. To create sculptural pieces

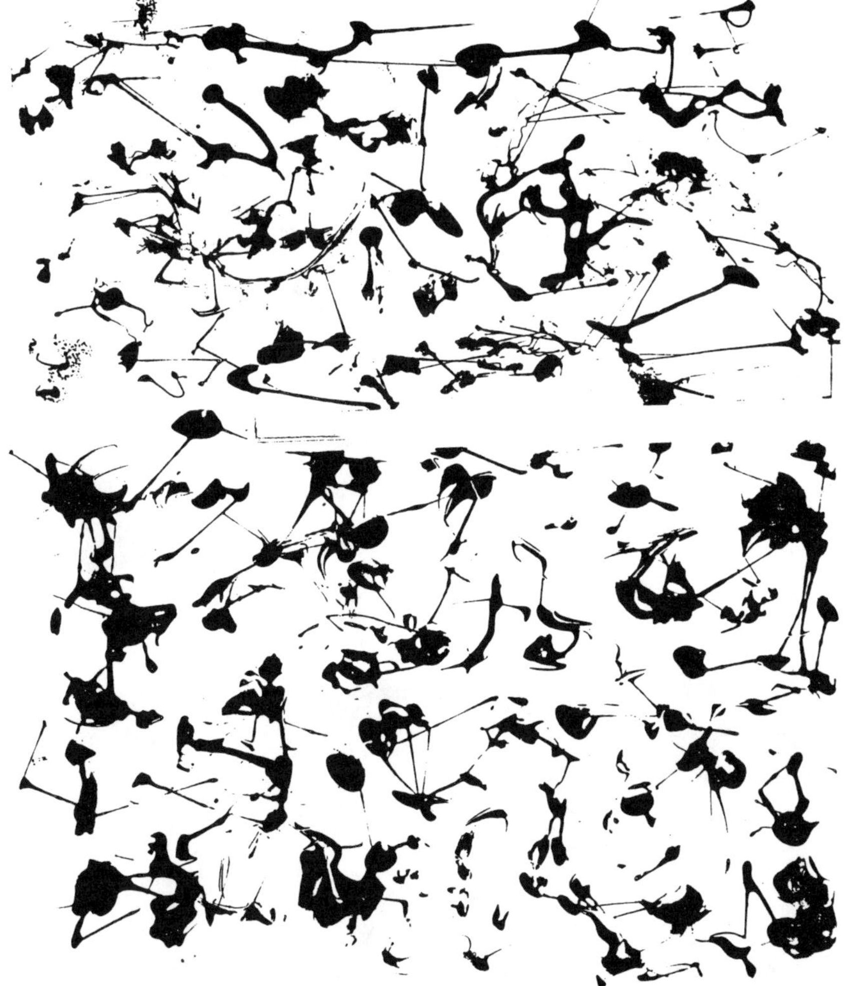

Fig. 35. Randomly created images painted on plastic film with a miniature spatula or needle dipped in carpenter's glue to which India ink has first been added. Having considerable depth in their original form, these images may serve as a starting point for the development of sculptural ideas by adding molten plastic where necessary or by cutting away superfluous or unwanted parts of the original. Instead of being used directly, they may be used as ideas or suggestions from which independent sculptural forms can be developed (see plates 1-5).

based on these would be an easy and pleasurable assignment, the thermoplastic material either melted directly onto the image and built up to the desired

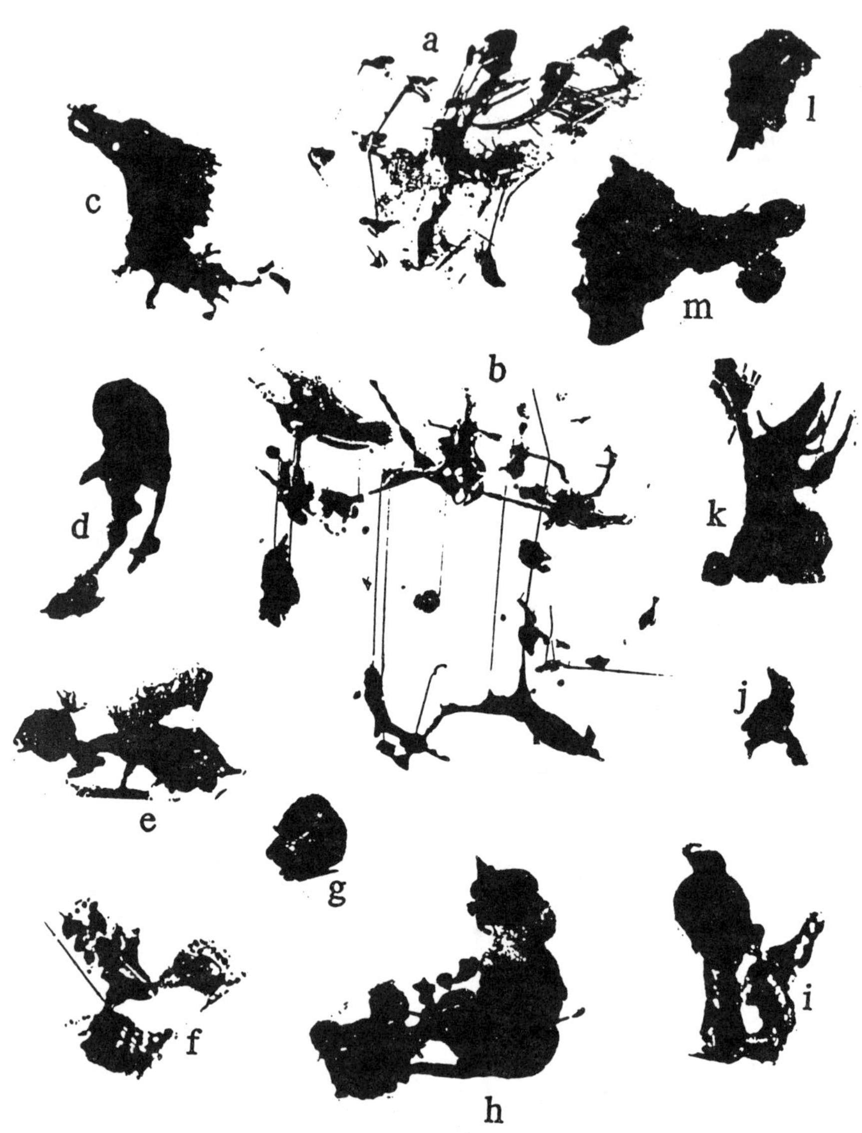

Fig. 36. Randomly created images produced in ink on drawing paper, using a miniature spatula.

thickness of a three-dimensional form or simply built up freehand, with or without an armature, as the structure of the particular form dictates.

The images shown in Fig. 37 were basically fortuitous, though slight retouching was employed to strengthen the features of b and c. The conventional Rorshach technique of producing bilaterally symmetrical ink blots (as in a) is a simple way of creating ideas translatable into scenic backdrops for sculptural miniatures or into actual sculptural pieces themselves.

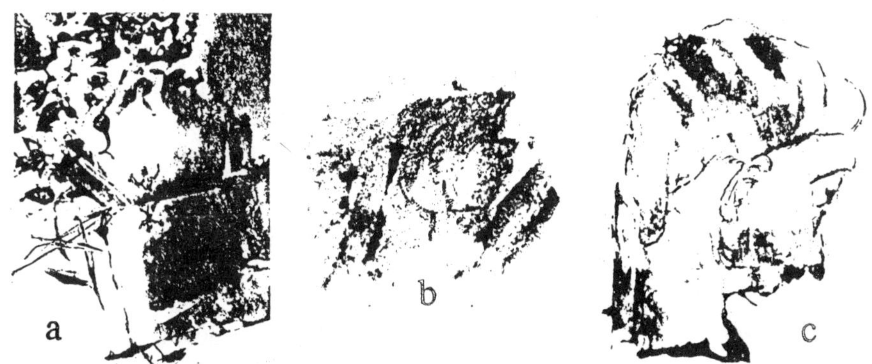

Fig. 37. **Images created fortuitously** in non-sculptural projects. a. Conventional ink-blot technique with folded paper. b. Residue left after cleaning glue and sawdust from a small spatula on waste cardboard. c. Image left after cleaning spatula. Slight retouching to emphasize facial features in b and c.

Various other ways of creating backgrounds against which to view the sculptural pieces suggest themselves. Striking ones, resembling woodcuts, may be made on scraper board, for example, but if one wishes to be a purist and use only thermoplastic materials, a substitute for scraperboard can be improvised by painting a piece of opaque white polyethylene black and then scraping away parts of the dried paint to produce a design. Dilute acrylic gesso can be spread on a glass plate, smudged with a brush and left to dry. The dried film can be viewed with transmitted or reflected illumination, the effect with one often different from (and sometimes superior to) that of the other. The design can be altered and developed further by scraping, by adding material, and by blending with polyethylene (using the electric-loop tool). An alternative method of creating imagination-stimulating forms is to press a bit of acrylic gesso (with carpenter's glue and India ink added if desired) between two glass plates and then pull these part, leaving film in fascinating shapes on each plate. A piece of hard-surfaced paper pressed against gesso on a glass plate and then pulled away can leave challenging patterns on both surfaces, the ones on the glass plate effectively viewed when a strong light is shone on them from the back. Of course, many of the conventional methods of printing, embossing, and transferring designs may be employed in ways that will stimulate the imagination of a miniaturist in search of creative sculptural ideas.

The aleatory aspect of various technical procedures

The element of chance which enters some of the procedures given above and elsewhere in this book can be used to great advantage in stimulating one's imagination as one attempts to produce satisfying sculptural pieces. The pulled-adhesion procedure, for example, whether one is working with molten thermoplastics sandwiched between wooden blocks or with gesso between glass plates, often gives unexpected results, producing forms that seem to have arisen by chance alone, although it is quite apparent that the limits of variability are basically set by the interplay between various physico-chemical and mechanical factors that act upon the materials during the formative process. This interplay gives rise to an excitingly broad range of forms with which one's imagination can play in its search for challenging and satisfying compositional ideas. In the same way, the procedure for producing anlagen by pulling a molten mass of plastic in various ways with such tools as pliers, needles, and tweezers enables one to develope forms in a quasi-random way that may be sufficiently suggestive of an ultimate goal to give—in a sense, automatically—direction to the sculptor (compare here Lipchitz's semi-automatic sculpture). Random forces are allowed to act to a point, at which the sculptor intervenes in a calculated and planned way, leading (either consciously or unconsciously) to the germination or development of forms unforeseen. Before the point of conscious or calculated intervention is reached, however, the course of development may have been defined in only the most open and loose, or free, manner, giving relatively free range to one's imagination, but by the time this critical stage is reached, the imagination is being channeled along courses set by the deformation of the urschleim through various procedures, and the direction to be taken becomes clearer and more rigidly defined henceforward and until the piece is completed. Chance, in the form of a not-so-easily-predictable interplay of numerous factors (some quite subtle, others less so), thus can play an important and valuable role in the creation and development of basic ideas for miniature sculptural forms.

The relation between thermoplastic properties and artistic style

The thermoplastic materials recommended early in this book, if worked with an appropriate range of procedures, are capable of responding to a wide variety of stylistic approaches. If suitably compounded, and if stock pieces of appropriate size and shape are selected, they can readily be transformed into anlagen from which the artisan, using the appropriate combinations of both classical sculptural and modelling techniques, can develope finished pieces having almost any conceivable form, although, of course, some are more easily achieved than others.

Few materials—natural or synthetic—combine so satisfactorily for the miniaturist the attributes of economy, ready availability, easy workability, and the potential for achieving expressive variety and aesthetic satisfaction as do these thermoplastics. In few other materials is it so easy to achieve the dynamism of the Futurist, the vitalism of Modern Sculpture, the wit or satire of the Surrealist, the fluidity and organicism of the Impressionist, or the humanism and cohesiveness of the Classicist. Even the synthesis of energy and mass achieved by the Cubists with their "scaffolding of coalescing planes" can be achieved, though, admittedly, with a bit more difficulty than is encountered in producing the more cursive natural forms of the Realist or the Classicist. In this technically more demanding style, the harsher, light-exploiting surfaces may be produced in several different ways: by direct modelling or buildup in much the manner of the Constructivists, by blow-molding component elements over modular or non-modular patterns for subsequent assembly and buildup, and by various types of press molding, die casting, and stamping from patterns initially produced in the same plastic material or in any one of a wide range of materials, from paper and wood to wax, clay, and metal.

Part of the mystique of these thermoplastic materials is that they respond so satisfyingly and so easily to heat. In doing so they behave in such rigid conformity with the physico-chemical laws of nature that they naturally produce, or, with artistic encouragement from the sculptor, can easily be made to produce forms that meet the qualifications of the internal harmony found in the growth of organic forms. Thermoplastically-produced forms have an organic quality about them that lends itself readily to the imitation of naturally occuring ones, both organic (in the biological sense) and inorganic (in the physical and non-chemical sense). One is easily led to speculate that this predilection for organic (in the biological sense) morphological harmony is not merely a reflection of the organic (in the chemical sense) material itself. Just as the harsher form of inorganic (i.e., non-living) materials in their pristine crystalline state (of which much Cubist, and much of the more linear types of modern art, seem to be reflective) is moderated and made more cursive and gentle in outline by the erosional weathering forces of nature, so are these plastics—through the application of heat—easily turned into such "organic" forms as the "concretions" of Jean Arp, the garden sculptures of Barbara Hepworth, and the monumental sculptural works of Henry Moore, most of which look so much as if they have been rounded by the abrasive action of the waves, stones, and sand of some tumultuous rocky shore.

The plastics used in making the miniature sculptural pieces illustrated in this book uniquely combine many of the properties of the materials most commonly used in conventional (i.e., *subtractive*) sculpture, whether classical or modern. In their unheated state they can be carved like stone, wood, clay, or wax, although the easiest way of achieving this is through the use of the electric-loop tool, which, in effect, takes full advantage of the thermo-*plasticity* (used here in its physico-chemical, rather than its more rigid artistic sense), a property lacking in stone (unless it be raised to temperatures not readily or practicably achieved outside an experimental laboratory) and wood, but present in wax and metals. In the general area of pure modelling (i.e., *additive* or *built-up* sculpture), from which stone and wood have classically been excluded (though in modern times these, too, can be used—as metals frequently are—to build up sculptural pieces), the thermoplastics can be employed most effectively when used more or less as clay, wax, and metals are, the secret, again, being to use heat, as in modelling wax or metal. Clay is generally more easily worked at room temperatures and with unheated tools than are the synthetic plastics, but thermoplastics work, based principally on the carefully controlled application of heat, applied either directly or by means of heated tools (as with wax working, which it most closely resembles), can expand easily and vastly the horizons of one's creative sculptural designs.

The modern fascination with the various properties of metals in either their cold or heated state can justifiably be extended to the thermoplastics, which to a remarkable extent exhibit most of the characteristics to some degree and in ways which generally make their realization achievable much more easily and far less expensively. With the application of properly controlled heat, a suitably compounded thermoplastic can, like a metal, be drawn out to produce rods or "wire" and spun out into the finest of filaments, or it can be press worked or cast with suitable molds, dies and stamps, paralleling most of the operations effectively used with metals in industry but some of which are not as yet so widely applied in sculpture. The ease with which the thermoplastics can be welded, using the electric-loop tool and extruded rod stock in a manner closely analogous to that of the metal worker, places at the fingertips of the sculptor most of the creative magic of the constructivists and other modellers who rely so heavily on metals in achieving the built-up compositions that are so conspicuous in the modern scene.

Some of the working properties and behavioral characteristics of molten thermoplastics are sufficiently similar to those of glass that the sculptural miniaturist would do well to investigate and apply some of the glassworker's technique to these materials. The graceful, flowing curves and rounded surfaces resulting from the response of molten glass to the complex of forces applied by the artisan and by nature are easily duplicated when working with thermoplastics, the cursive quality of blown glass much more easily reproduced in these than is the linearity and angularity of some modern sculpture.

The wide range of working properties of thermoplastics is conducive to a commensurately free eclecticism in the styles one wishes to employ, this freedom possibly exceeding that in the use of other, more conventional sculptural materials. Whether it be to depict the stately cohesiveness of a mass in repose on a horizontal surface, the grace and elegance of smoothly blended surfaces, the dignity of a classical torso, or the melting watches of a Dalí landscape (see Plate 13,c, for a sculptural pun on this), it all can be done thermoplastically. By extension, in moments of espieglerie, when one's more lighthearted flights of fancy desire expression through the soaring dynamism of cursive lines swept into elaborate arabesques, this, too, can be achieved with ease and grace in these near-magical thermoplastic materials.

The challenge of seeking and exploiting relationships between music and sculpture, as Kandinsky did with music and painting, led to the inclusion of a musical appendix in the present work, the hope being that the reader might be inspired by it to explore the relationship in creative ways of his or her own design.

14

ANGST AND OPTIMISM IN THERMOPLASTIC MINIATURES

IT IS NOT inconceivable that at least some of the *Angst* encountered so widely in modern art and sculpture and so often attributed to an alienation from modern technological society is, rather—at least in the case of true sculpture (carving), as opposed to modelling, and at least in the case of the "New Iron Age" sculptor and modeller attempting to apply the same disciplines and procedures to metal—a reaction to both the physical and financial arduousness of the task of carving natural stone. Perhaps miniature thermoplastics sculpture is a way of avoiding some of the stress that leads to *Angst* and can open new, optimistic vistas for the future. Certainly it seems likely that Brancusi would not have encountered the same difficulty with the U.S. Customs authorities if his *Bird in Space* had been a plastics miniature less than two inches high instead of a large hunk of polished bronze, nor would Barbara Hepworth have had to tramp through inches of millstone grit and chips during her visit to his studio if his material had been a few discarded high-density-polyethylene yoghurt containers. Moreover, I dare say that any thoughtful sculptor working at thermoplastic miniatures runs the risk of being so carried away by the challenges and satisfactions inherent in the material's mystique as to be imbued more with *Hope* than with *Anxiety* in creating a finished work, in vindication of Werner Haftmann's observation (Evers, 1963; Moore, 1964, p. 225) that to consider only the *Angst* in modern sculpture is to ignore the very real elements of hope inherent in so much of it. The mere technical challenge of attempting with thermoplastics and on a miniature scale to interpret and depict in a surrealistic way some of the cultural complexities of modern society can help the would-be sculptural miniaturist to throw off the shackles of modern pessimism and, like the originators of Surrealism themselves (and to paraphrase Stich, 1990) reclaim "some of the suppressed realism of human expression" while "confronting contradiction, difference, disjunction, multiplicity, rupture, and incongruity".

An important part of the numen of thermoplastics miniatures is its hopeful (although itself to a degree aleatory) element, the verdant promise deriving from their environmental friendliness. Spatially, they ask little of nature. Even the few ounces of scrap polyethylene salvaged from a plastic

130

container destined for the garbage dump can remove a small additional burden from the environmental clutter and pollution when it is converted into a work of art to be treasured for generations (at least by the sculptor's own family, if not the world at large), in a sense a miniature monument to the technological and artistic ingenuity of modern society. No tree was sacrificed for its wood, no landscape scarred by a quarry face and waste heap for its stone, and but an infinitesmal fraction of the energy required to produce Brancusi's *Bird in Space* was needed to produce the material for a lifetime output of plastic miniatures of the type being considered in this book.

The optimism expressed so effectively in the poetic metaphors of Marinetti's 1909 manifesto (in *Le Figaro*, Feb. 20, 1909) for the Futurists—his encomium to a modern technological world, with its "factories suspended from the clouds by... strips of smoke" and "its bridges leaping like gymnasts over the diabolical cutlery of sunbathed rivers"—finds up-to-date (albeit meiotic) realization in miniature thermoplastics sculpture. It is not unlikely that Marinetti (the son of a wealthy industrialist), who thought that "a racing motor car, which looks as though running on shrapnel, is more beautiful than the *Victory of Samothrace"*, would appreciate the miniature pun on the Holy See (Plate 13, a) achieved with a bit of plastic scrap, or might even have smiled wanly at the little motorcyclist and his clouds of smoke (Plate 5, i) suggested by one of the glue blots mentioned elsewhere in the text. It seems almost certain that he would have seen and apotheosized in rousing acatalectic hexameter the grandeur in the autoregulatory realignment of the sliding molecules in a molten blob of high-density polyethylene transmorgrified, through the Lilliputian conflagration of a luminous Satanic mill of a superenergized nichrome wire, into a microscopic distillation of the Futurist's own perfervid numen. But, while to move from such a Cubist-influenced work as Brancusi's polished-bronze *Bird In Space* to a miniature replica in polyethylene was relatively easy, I learned, from bitter experience, that to accomplish the same with Boccioni's *Unique Forms of Continuity in Space* or his *Anti-Graceful (The Artists' Mother)* is a most demanding assignment, not realizing at the time the true significance of Boccioni's own warning that even the primitiveness of the Futurist "is the extreme climax of complexity".

The Musical Appendix to the present work is intended to exemplify the optimism in modern music and the ease of relating this optimism to that in thermoplastics miniatures, a world (albeit miniature) of possibilities existing there for dramatic and choreographic interpretation through use of the miniature shadow-play theater shown in Fig. 42, a.

SCULPTURE FOR THE MASSES

The sculptor who does his work in stone
Has life-long fears of silicosis.
The dust he makes we can't condone,
But perhaps it helps his halitosis.

The floor of one who works in wood
Is inches deep in chips, of course,
Except where last he stood.
His fate, alas, is thus perforce.

If teeth and tusks are what you require,
Procurement is now a task,
For the ivory that we all admire
Is no longer there when we ask.

All types of metal abound today,
If you like to work in such stuff.
There must be those who find it play,
But I consider it tough.

An intermediary fine
Is one of the many waxes,
But for ways of making it last we pine,
And this our brain o'ertaxes.

Of course, we still can turn to clay
To cast in bronze or to fire,
But for pieces small there must be a way
More easily our ends to acquire.

All this above is why I propose
To work with plastic waste.
Some may say it lacks class, I suppose,
But then it's a matter of taste.

(cont. p.133)

(cont. from p. 132)

If you find that your taste, like mine,
In art and music and such
Is something you'd like to refine,
From plastics you stand to gain much.

They're easy to work, the cost is low,
The range of expression is great;
Also, you can take them wherever you go
And work on them there at full spate.

Sculpture ideal as an art for the masses,
Pieces galore wherever you roam;
No need to go out at night for your classes,
You learn it yourself at home.

So, I suggest you give them a try,
The investment you'll make is quite small.
Don't be surprised if soon you see why
Under their spell it's so easy to fall.

PART FIVE

STORING, VIEWING AND DISPLAYING MINIATURE SCULPTURAL PIECES

15

STORING

MANY TYPES OF commercially-produced containers, particularly transparent plastic ones, may be obtained to accommodate the miniature sculptural piece, but one can experience some difficulty in locating a source for these, particularly in the relatively small quantity one requires. It is more economical, of course, to make one's own from household waste of various types.

If one prefers to make one's own containers, several ways of doing so come to mind. A satisfactory home for relatively large pieces may be made from the cardboard containers used for frozen juices, but, as will be discussed later, pill bottles, when cut into segments of appropriate height, are particularly suitable for the smaller pieces. For most of my larger pieces, the 6-ounce "can" is suitable, but some of the largest ones require the 12-ounce size. Cut the container to a suitable height for the piece it must accommodate, the usable section to be that with the metal bottom permanently attached. It is easy to make the cut with a band saw, but *take the precaution of using a rip fence and making a special tool (a length of sturdy wooden rod stock inserted into the can and pressed against the bottom to hold it against the rip fence) for holding the can securely in place against the blade while the cut is being made, so that it cannot be jerked away and thrown from the saw. The procedure can be dangerous, so be very careful when attempting it.)* After cutting, the can, inverted, so that the bottom becomes the top of the protective case, may be left without further modification if one is content to have an opaque cover and relies on an external label to identify the contained sculptural piece, but if one wishes to be able to identify and examine the piece without removing the cover of the storage container, a window must be added. This is done by cutting away part of the container and, for swank, fastening over this a clear piece of plastic from your scrap box. The cutting may be done with a knife or chisel, preferably with the container supported on a wooden mandrel held in a vise, or it may be done with a small circular-saw blade fitted to such a tool as the Dremel Mototool. *Considerable care, however, must be exercised when doing this, because the saw can easily and quickly snatch the container from your hand, possibly cutting your fingers in the process.* I prefer to use a craft knife for the job, using the

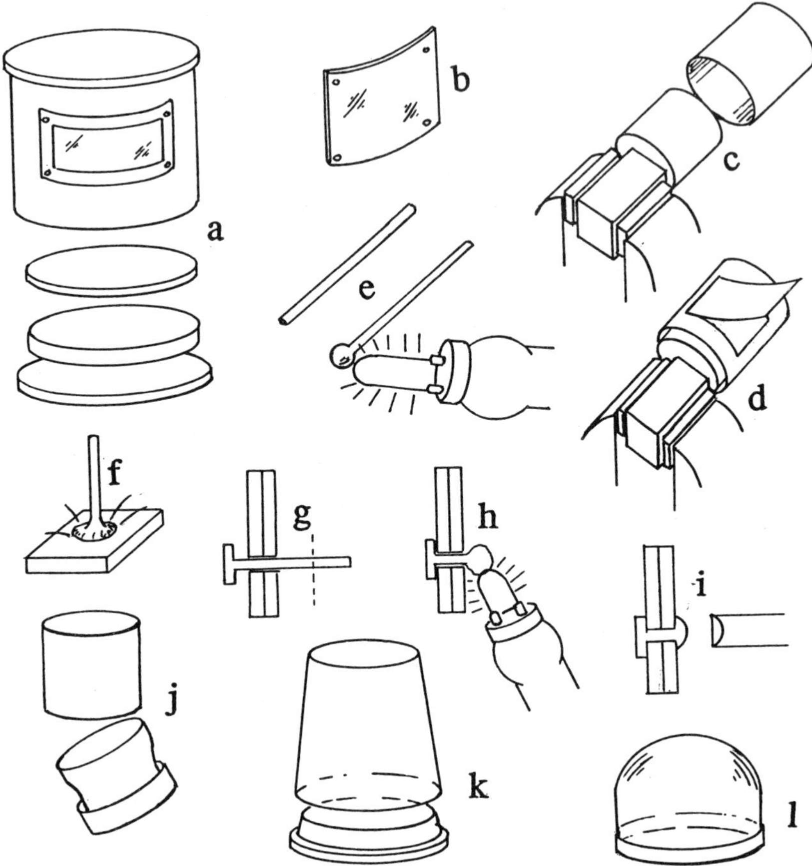

Fig. 38. Storage cases for miniature sculptural pieces. a. Exploded view of case (above) cut from a cardboard frozen-juice"can" fitted with a plastic window and (below) the three components of the mounting-board base for the sculptural piece itself. b. Plastic window with holes drilled for attaching it (with plastic rivets) to the case itself. c. Mandrel (held in vise) on which the opening for the window is cut (see *d*). e-i. Stages in forming and installing a plastic rivet to hold window in place (e). Plastic rod stock is heat-softened at one end, and this is flattened (f) against a cold surface. g. The rivet is inserted through holes in the case and window and trimmed, leaving stock for melting (h) and final shaping (i) with a cupped tool. j-l. Other types of cases: j. Segment cut from a pill or medicine bottle to desired height and fitted with base (and window, if desired). k. Plastic tumbler for larger pieces (fitted with a base, as in *a*). l. Commercially-produced display case of the type obtained from candy/novelty dispensers.

mandrel/vise setup shown in Fig. 38, c-d. Use the knife carefully to avoid cutting yourself. Once the cut is finished and the severed piece removed, the edges of the opening thus produced should be smoothed by sanding or filing.

The window for the cover may be cut from soft-drink bottles or transparent sheet stock from the scrap box, one advantage of using bottles being that the curvature of the molded plastic (if a bottle of appropriate size is chosen) conforms roughly to that of the cover and can more easily be fitted to it, not requiring reshaping (by heating on a mandrel of proper diameter).

The plastic cover for the window area may be attached with small rivets made from scrap polystyrene sprues or from rods of high-density polyethylene extruded as described in an earlier chapter. The rivets are made by heating one end and flattening it against a cold surface to form a head (Fig. 38, f-i), then cut to the required length, leaving enough for flattening. The rivet is then pressed (from inside the container) through holes drilled for it in both the clear plastic sheet and the wall of the container, after which the free (outer) end is softened with the electric-loop tool and flattened with the blade of a screwdriver, putty knife, or a piece of metal scrap. When applying the rivets, it is helpful to hold the container over a fixed mandrel (held in a vise, for example), this to keep the rivet in place while the free (outer) end is heated and flattened, enabling one easily to achieve a tight fastening. If one wished to finish the exposed end of the rivet in a professional manner, a tool having a hemispherical cavity (Fig. 38, i) may be made and used in the flattening process.

The basal unit of the protective case, separate from the juice-can component, consists of a disk of 0.25-inch Masonite (or equivalent) to which a slightly larger disk of posterboard or mounting board is glued, the latter sufficiently large to form a shelf against which the cover will rest, leaving a lip to facilitate opening the container. Both the disks may be cut (fly cutter and gasket cutter, respectively) in the drill press. It is important that the disks fit snugly into the cardboard container, so that the basal unit does not fall out easily or accidentally but must be removed by a conscious effort. When the sculptural piece has been completed, its own mounting plate should be glued to the top of the Masonite disk, forming a unit which can be handled as such when the container is to be closed or opened and the sculpture removed.

Using the same principle as that for the protective case just described, i.e., a salvaged ready-made container attached to a homemade basal unit, one can fabricate cases in considerable variety and accommodate a wide range of sculptural pieces. Thus, for example, segments cut from soft-drink bottles and fitted with bases like those just described will accommodate larger sculptures than those considered above, while ones made from opaque or transparent plastic vials and pill bottles can be used for other, smaller ones. Cases may be improvised from plastic drinking cups (Fig. 38, k) for larger sculptural pieces, and once the search for suitable scrap is launched, one will undoubtedly find various pre-formed objects that can usefully be converted or adapted for one's needs.

An alternative way of storing the sculptural pieces is suggested in Fig. 39, a. Here a two-quart milk carton has been modified (top section cut away) to leave an open-ended container into which can be inserted a storage tier, consisting of posterboard plates separated by short lengths of dowelling, the unit removable as such. Each shelf in the tier accepts four sculptural pieces on their 1.75-inch basal disks, the unit accommodating a total of 20 pieces.

A simple but effective storage/display case (Fig. 39, d) can be made from the pull-tab of a pop-lid can by binding it with a strip of card stock to a posterboard backdrop, the depth of the case determined by the width of the binding strip. The strip is glued (hot-melt glue) inside the turned-over edges of the pull tab and to the edge of a facing piece of posterboard stock glued to the backdrop. In selecting pull-tabs to be used for these cases, choose ones having a channel on the back (formed by the rolled edges of the metal) wide enough to admit the strip of card stock for secure gluing. Run a bead of hot-melt glue into this channel, heat the metal to soften the glue, press the strip into the glued channel, and allow to cool before use. The sculptural piece to be displayed in the case should be mounted by suitable support rods (extruded rod stock is good for this) to the face of the backdrop before the binding strip (which forms the side, top, and bottom walls of the case) is glued in place.

Smaller sculptural pieces also can be displayed effectively and stored safely when mounted inside the plastic cap of a soft-drink bottle. A posterboard disk of appropriate color is first cut (with an arch punch) and inserted against the bottom of the cap to provide suitable contrast to the sculpture, and a protective cover is easily thermoformed by the yoke/plug process illustrated in Figs. 23, a and 40, f , such a cover shown in Fig. 40, e.

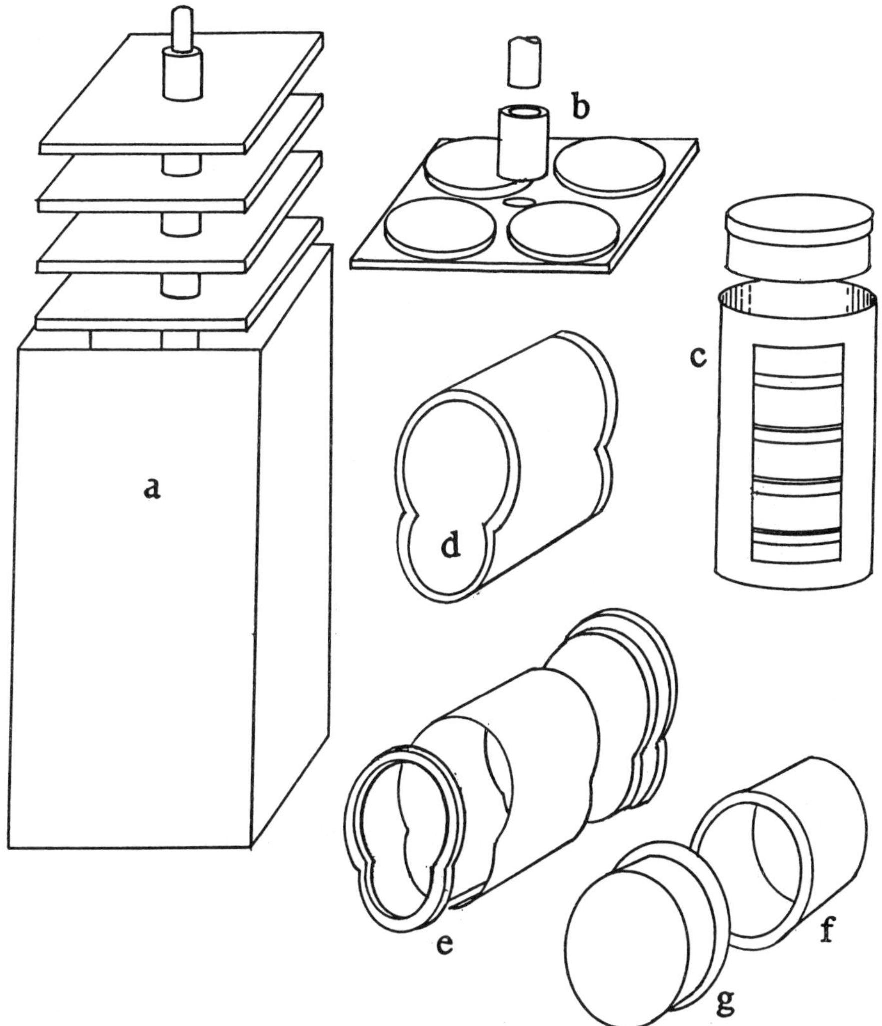

Fig. 39. Display/storage containers for miniature sculptural pieces. a. Storage container made from a two-quart milk carton. b. Exploded view of one shelf of the removable rack from the storage container shown in a. Circles indicate the spacing of the four sculptural pieces accommodated by the shelf; also shown is the spacer and a segment of the dowelling that forms the central support of the rack of shelves. c. Storage container for a stack of plastic boxes containing sculptural pieces; container made by cutting a vertical viewing slot in the front of a cardboard fruit-juice "can". d. Display/storage case framed by the pull-tab from a pop-top can lid. e. Components of the case shown in d. f. A small display/storage case made from the cap of a soft-drink bottle fitted with a thermoformed cover (g).

16

VIEWING AND DISPLAYING

Simple viewing devices

BECAUSE MOST OF the sculptural pieces, particularly those made without armatures, are quite delicate and easily damaged, steps should be taken to protect them when they are presented either for informal viewing by others or in formal displays. Small transparent plastic containers of various types may be purchased, but suitable ones may also be made from the plastic scrap that accumulates in various forms around the house. If the containers are to be handled by viewers, the sculptural pieces should be securely attached to the container and steps should be taken to prevent opening, although, of course, they are seen to best advantage when not encased. If time permits, the sculptor can uncase each piece in turn and show it to the viewers, but even so, accidents occur and pieces can be damaged, particularly in the excitement of a crowded room.

Specimens stored in opaque cases or even in those with plastic windows (these in some cases are more an aid to quick identification of the piece than for serious viewing) must be removed from their case for viewing and display, in which condition they are, of course, highly vulnerable unless placed inside a conventional display cabinet, where they can be seen but not handled.

A display device intermediate between the uncovered piece, completely without protection, and a proper display case is the display frame shown in Fig. 40, e. The device consists of a specimen chamber held in a bent-acrylic frame (by screw cap; both chamber and cap obtained from a pill bottle), an orifice in the frame accepting the screw section (neck) of the bottle, which is held in place by screwing the cap onto it.

From a stock of discarded pill bottles of various sizes (sweet are the uses of adversity), one can easily make a number of the chambers in which sculptural pieces can be mounted for permanent storage or for immediate display. To make the chamber, the neck section is removed with enough of the body proper to provide comfortable housing for the mounted sculptural piece. If the cut is made on a lathe operated at slow speed, a sharp knife the cutting tool, a single cut suffices to separate the neck section and leave

140

enough of the bulk of the body section for use as a permanent storage case for other sculptural pieces. To make the cut, the bottom section of the bottle should be held in the 3-jaw chuck, and a live center should be used in the tail stock *to prevent accidental dislodgement and its attendant dangers.* Some experimentation may be necessary to determine the depth preferred for a particular chamber, but in most cases some variation is not objectionable. The opaque brown bottles sometimes encountered make particularly effective chambers when used with white or ivory sculptural pieces, but, of course, one can paint the segments of white bottles to obtain the desired contrast. A colored disk of posterboard inserted securely behind the screw cap and the neck of the bottle makes a pleasing background against which to view the piece, the piece itself attached inside the frame by lightly fusing it to the neck with hot-melt glue and HDPE. When thus mounted, the piece is reasonably secure unless it is poked with a finger intent on destruction or damage, but it can be viewed from the front in an effective setting formed by the salvaged bottle remnant and the colored background-disk.

As shown in the illustrations, the cap of the bottle is saved and used to affix the neck section to the plastic frame in which it is mounted. Some advantage accrues from using bottles having a safety cap, because the larger cap is easier to hold for viewing the piece, is less likely to work itself loose during handling, and is not easily removed except by a determined effort, although it can be removed when this is necessary, as, for example, when changing the background disk or working further on the piece itself.

A vacuum-formed or yoke/plug-molded polyethylene cover for the open face of the display chamber provides additional security for pieces not being viewed. To make such a cover by vacuum forming, place a wooden disk just slightly larger than the body of the pill bottle onto one of the support plates for the tool shown in Fig. 22, c and pull a heat-softened blank of polyethylene over it; allow to cool and trim away the excess material. A similar disk can be used to yoke/plug-mold a cover (Fig. 40, f). Add a shaft that can be chucked in the drill press, and prepare a yoke (from Masonite) through which the plug and the heat-softened polyethylene (of a blank clamped securely between retaining rings) can be pressed to produce a cover of the desired depth.

The sculptural piece in its display chamber may safely be stored simply by inverting it onto a flat surface. As one's collection grows, a simple cabinet having shallow, tray-like drawers will be found to be a worthwhile home-workshop project, permitting one to store large numbers of pieces in relatively little space safely and efficiently.

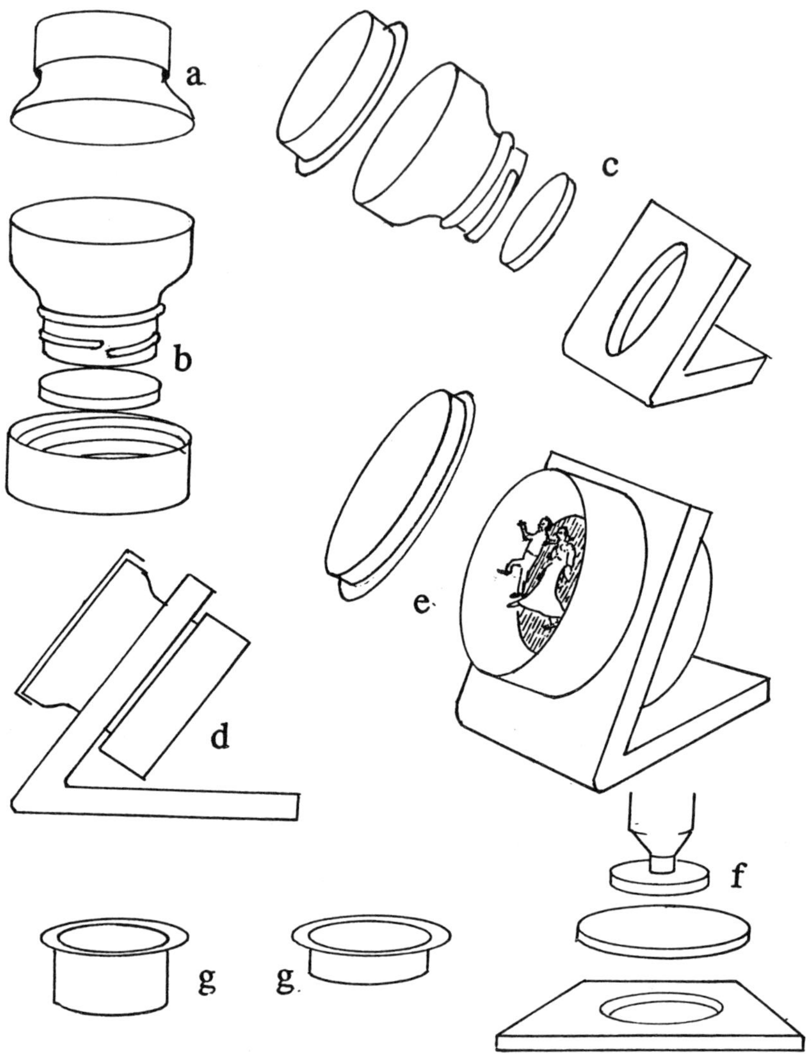

Fig. 40. Display device for individual sculptural pieces. a. Neck section of pill bottle (with its screw cap) cut from body, ready for use. b. Neck section (above), mounting disk (middle), and screw cap (below), ready for insertion into heat-bent acrylic frame. c. Exploded view of the device. d. Sectional view of the components of the display assembly. e. Assembled display (protective cover open). f. Procedure for forming a protective cover (polyethylene) by the yoke/plug molding process: a wooden disk (plug) used to press a heat-softened polyethylene blank into a Masonite yoke to a depth sufficient to hold the cover securely to the display chamber. g. Yoke/plug molded covers (caps) for display devices.

While the display chambers with their encased sculptural pieces can, with a word of caution to the viewers, be circulated with reasonable safety among small audiences, it often is best to provide individual frames or stands which will present the piece and its container at a convenient viewing angle in such a way that no handling by the viewer is required. The frame shown in Fig. 40, e holds the display chamber securely when the screw cap is attached to the threaded section of the severed pill-bottle section.

A more elaborate way of presenting pieces for viewing is by means of the viewer shown in Fig. 41, a. A simple plywood box fitted with a hinged lid, a rotatable specimen holder (stage), and either a magnifying or a reducing lens (the two interchangeable), this viewer is loaded and unloaded by the sculptor and passed to each viewer in turn. The shaft of the specimen platform can be slid upward through a simple sleeve-bearing for loading and unloading, and it may be rotated manually to permit viewing all sides. The lid is left open during the viewing process so that light falls on the subject. The lens holder slides onto the viewing tube and is held in place by two press-fitted rivets, these easily removed with the fingers when the lens is to be changed. A vertical slot on each of the two rear corners of the box accepts colored pieces of poster board, a simple way of changing the backdrop to provide best contrast and viewing conditions.

Carousel viewer

The most elaborate of all the viewing devices so far considered is the one shown in Fig. 41, b, a miniature carousel gallery. Easy to carry and quickly set up or dismantled, this viewer gives maximum protection to the sculptural pieces and provides the most effective viewing conditions. It may easily be assembled in the home workshop in a few to several hours. The unit, consisting basically of a cookie tin, a tuna can, and a large plastic pill vial, is fitted with a specimen-carrying platen and interchangeable reducing and enlarging lenses; it is driven by a 1/2 r.p.m synchronous motor and is supported atop an adjustable photographic tripod, which permits one to adjust its height to a comfortable position for each viewer in turn. A dozen or so sculptural pieces are loaded in advance onto a circular plywood or Masonite platen held in place by two locator pins on the substage platen. A single background may be provided for all the pieces by inserting a ring of colored posterboard toward the center of the platen, the ring held in place with glue tabs or by screws; a more effective approach is to provide individual slots for colored cards appropriate to each of the pieces, the cards easily changed to match the particular piece for which they form a backdrop.

A small light attached over a hole cut in the wall of the cookie tin illuminates the interior, a microswitch permits the viewer to stop the rotation

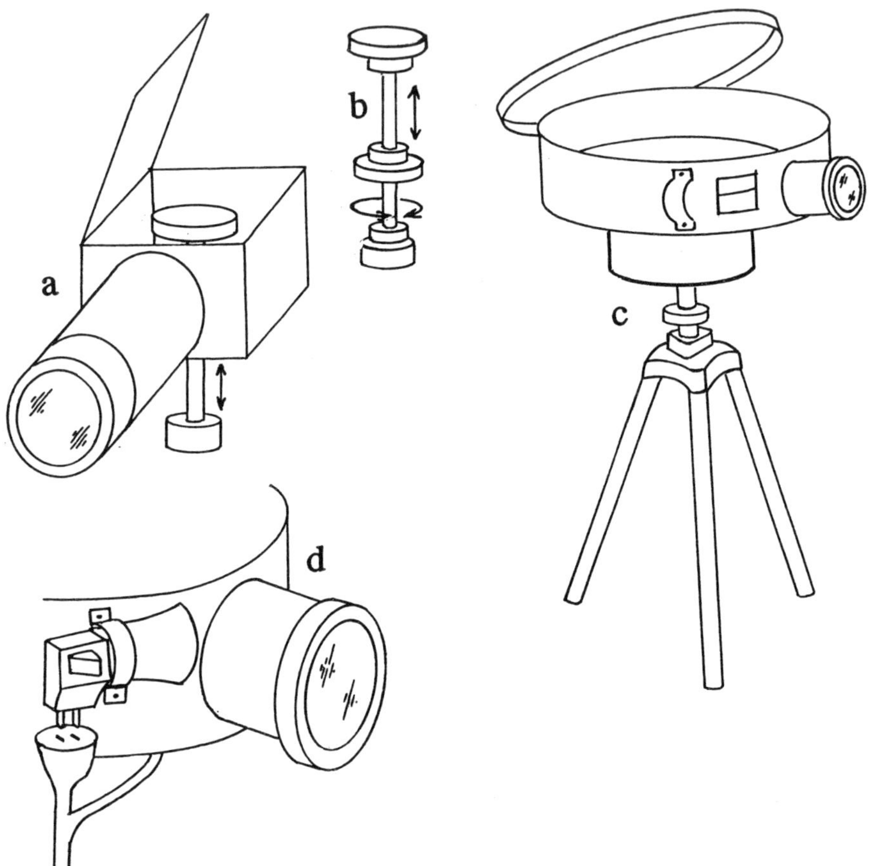

Fig. 41. A simple viewer and a carousel gallery for miniature sculptural pieces. a. Homemade viewer for miniature sculptural pieces. A heavy cardboard tube, fitted interchangeably with a reducing or an enlarging lens, is attached to a hinge-top box (wooden), and the sculpture is attached (wax) to a rotatable and vertically adjustable viewing stage (b) which slides up from the bottom of the box for viewing or for loading and unloading. c. An assembled carousel gallery mounted atop its photographic tripod and ready for use, the cover ajar to show the interior of the viewing chamber. A microswitch enables one to start or stop the rotation of the specimen platen at will. d. Detail of the illumination system, a small night-light clamped over an opening in the wall of the chamber.

when desired, and a music-box movement, activated by a "boredom relief" switch, adds zest to the entire presentation.

A shadow theater

A simple device for use in projecting and viewing shadows cast by miniature figures is shown in Fig. 42. Easily made from scrap or inexpensive materials, the viewer is not only an effective device for displaying static pieces in silhouette but it makes possible the exploitation of the vast world of shadow plays and puppetry on a miniature scale for one prepared to spend the time to make the viewer, props, and figures. With such a facility at your beck and call, you'll not have to look far for an appreciative audience—youthful or adult—and can easily draw viewers away from the TV and VCR to watch your amateur productions and, more importantly, possibly even have them set about making their own viewer and starting a rival production company.

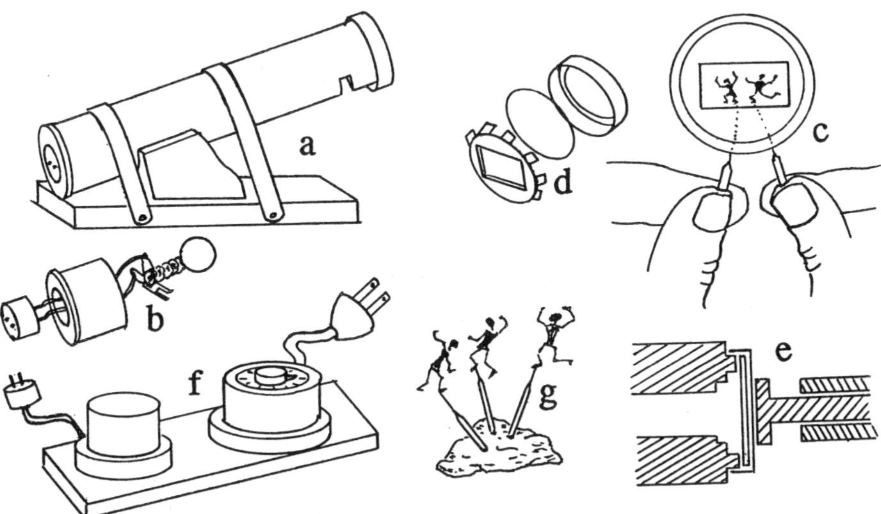

Fig. 42. A viewer for miniature sculptural pieces and shadow plays. a. Overall view of the device ready for use. b. Rear end of viewer, for attaching a penlight or a transformer-controlled bulb. c. View of screen with shadows of figures. d. Screen (i), retaining ring (ii), and postscenium or screen frame (iii). e. Safe method (in lathe) of cutting center from a jar lid to make retaining ring. f. Tranformer unit. g. Miniature figures mounted for use and stored on a blob of wax.

The body of the viewer is merely a heavy cardboard mailing tube (2.75" diam.,12" length) fitted at one end with a disk of tracing paper (translucent layout paper is excellent for this) that forms the viewing screen and at the other with a solid plug (wood, acrylic, etc.) drilled to accept and hold securely a small penlight or, as shown in the illustration (and preferably), a miniature bulb operated through a 6-volt filament transformer and a variable (0-120 volt Powerstat) transformer. The tube is attached to a wooden base by means of two straps cut from a length of steel binding tape salvaged from a packing crate. A hardwood quoin (wedge) pressed between the tube and the wooden base forces the former into a comfortable viewing position and holds it securely in place. A blob of wax (cheese wax is good for this), pressed onto the wooden base serves to hold the cast of characters in readiness for their entry (from below) onto the screen as the play progresses.

Unlike the stage of a conventional theatre, the viewing area of this stage (screen) is delineated by a shadow cast by a cardboard frame inserted just behind the screen, the conventional proscenium here, then, becoming a *postscenium*. An opening large enough to permit one to thrust props and characters up into the light path must be cut in the bottom of the tube, just behind the postscenium, and, if access to the path is required from other directions, other openings at appropriate points must be made (to permit floating clouds, birds and other flying creatures, for example, to put in their appearance when the script calls for it), care being taken, to avoid weakening the tube excessively, though, of course, strengthening strips of hardwood may be glued in place as necessary to compensate for such weakening.

The viewing screen is held in place across the front end of the body tube by means of a retaining ring cut from a mayonnaise-jar cap, the cutting best done on a lathe operated at slow speed and with great care, *the cap supported well both fore and aft, as suggested by Fig.42, e, the risk of having the tool dig into the metal and jerk the cap from the chuck considerable otherwise. By sandwiching the flat surface of the cap securely between a backing disk of Masonite and a support plate (faced with Teflon or other plastic of high lubricity) held against it in the tailstock, the risk of accident is reduced, but the operation should be performed only by someone experienced in the use of the lathe in such potentially dangerous procedures.* My scrap collection easily produced a jar lid that fit snugly onto the body tube of the viewer, but one could easily (and less elegantly) glue the screen directly to the tube, bypassing entirely the hazards just considered.

The light-holding plug (Fig. 42, b) at the opposite end of the tube may be turned from hardwood or acrylic, but mine is made from a small can filled with hard plaster (Fixall, from the hardware store) and capped with a disk of 0.125-inch Masonite. Though my viewer was initially made to accept and operate on a pocket flashlight (penlight), the advantage of having illumination of variable but controllable intensity soon became so apparent that the essentials for such a system were added. The bulb (a 6-volt miniature lamp with screw base) was screwed into a turned-brass housing that could be press-fitted into the plug at the rear of the tube. One lead from the 6-volt transformer was soldered directly to the housing. Contact with the center of the bulb's base was made with a brass rod press-fitted in a turned-acrylic plug which was cemented securely (epoxy) within the brass housing in such a way as to make good contact with the base of the bulb. The other lead from the transformer was soldered into a hold drilled for it in the exposed end of the brass rod. A generous coating of epoxy cement over the soldered areas finishes them off nicely.

Miniature figures to be used in animated productions are made on twisted-wire armatures to dimensions appropriate to the size of the viewing screen and are mounted (hot-melt glue) on a short length of fine wire attached to a short segment of wooden applicator stick or round toothpick. Several of the mounted figures may be kept handy for use by inserting the ends of their stick handles into a blob of wax (cheese wax useful for this) pressed onto the wooden base (Fig. 42, g).

The two transformers required for the system illustrated here are housed in homemade containers, the 6-volt one in a juice can (cardboard) cut to size and capped with its own lid (epoxied in place), the variable one in a housing fabricated from sheet metal and wood, although a tin can of proper size is also adequate and easily modified for the purpose. The encased transformers are designed to fit snugly into wooden retaining rings glued to a wooden base; the entire unit, merely plugged into the system when needed, can be unplugged and used elsewhere. Cases may be purchased for the transformers if desired, but the design possibilities for homemade ones are broader, the challenge of producing a pleasing design exciting, and the expense of making them delightfully inconsequential.

A nice challenge for one's model-making skills is a set of draw curtains for the miniature theatre. Because of diminutive size of the curtain assembly, however, this refinement (not essential to the operation, but highly effective if well executed) should probably not be undertaken without suitable tools (most of them described elsewhere in the book) and appropriate miniaturization skill.

The curtain assembly (Fig. 46) consists of :

    1. A mounting board to which the curtain rod and the eyebolts for the draw thread are attached.

    2. The curtains (2).

    3. A curtain rod which is pressed securely at each end into holes drilled for it in the mounting board.

    4. Brass rings (14) that slide freely over the rod, the curtains attached (hot-melt glue) to the rings at uniform intervals.

    5. Eye bolts (4-6) through which the control thread passes and by which its course is guided.

    6. Lead weights (14) attached (hot-melt glue) to the bottom of each pleat in the curtains.

    7. A control thread for opening and closing the curtains.

(Note that in the descriptions which follow, the indicated directions apply when the worker is in front of and facing the viewing end of the body tube.)

The mounting strip (Fig. 43, a), hardwood preferably, is drilled to receive the curtain rod in a press fit sufficiently tight to prevent accidental dislodgement during operation.

The curtains of my viewer were cut from cloth taken from a broken umbrella salvaged from a garbage can at a nearby shopping mall. Cut initially as a single unit (to be cut into two after rings and weights are attached), the cloth was first soaked well and then spread flat onto a small sheet of metal. It was then folded carefully and tightly around each of a series of metal strips (Fig. 43, b) cut to the width of the pleats to be formed in the curtains, the assembly of cloth and metal strips forming a tight stack which is then pressed firmly with a hot iron until the cloth is thoroughly dry and the

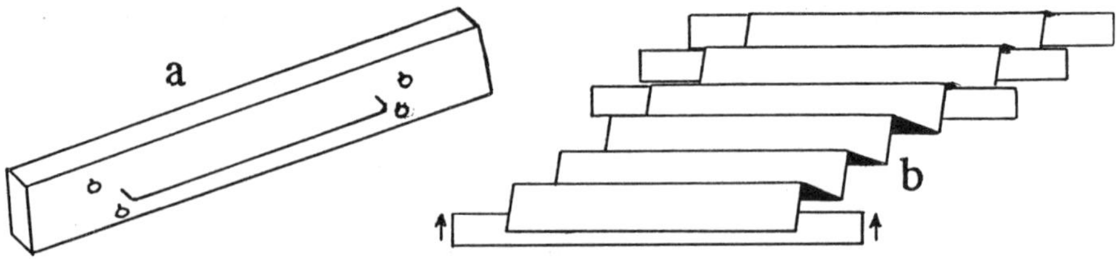

Fig 43. Curtain construction and mount. a. Mounting strip b. Curtain cloth folded around metal strips to form pleats to be set through the use of a hot iron.

folds of the individual pleats have been well formed. By cutting the metal strips long enough to extend past the ends (top and bottom) of the curtain cloth, it will be easy to clamp the assembled stack and hold the components in place during the ironing process. Well-formed pleats are essential to proper curtain action, so these must be made with great care.

The curtain rod is a length of brass rod (19-gauge) which will be bent at each end after the curtain is completed and it is ready for mounting.

The curtain rings are made (Fig. 44, a-f) by wrapping brass wire around a hardwood mandrel, cutting the resulting coil with a jeweller's saw to release each volution from the others, closing the resultant ring (slip it onto a tapered-tip mandrel at this point) with pliers, and soldering. To prevent having excess solder at the joint, it is wise to cut small pieces of solder in advance, making them of uniform size and just sufficient volume to produce a neat joint. A fine tip (0.625") should be used on a small soldering iron and the ring should be held securely on a mandrel (hardwood) during the soldering operation.The rings should be of a diameter that permits them to slide freely over the curtain rod, and care should be taken to clean each carefully after it is attached to the folds of the curtain, so that any excess glue used at this later stage is removed before the rings are slid onto the rod and readied for use. A ring is attached to the top of the leading, or center, edge of each pleat, and a thin rod of hot-melt glue (see description of this procedure on pp. 33-35) is used with the electric loop tool (in the fume hood, of course) to apply a small drop of glue that will bind cloth to ring securely.

The miniature eyebolts (Fig. 44, g) needed to guide the control thread for the curtains are made from fine brass wire twisted with one of the tools shown in Fig. 9. The eyes should be large enough to insure easy passage for the control thread, with no binding and with minimal friction.

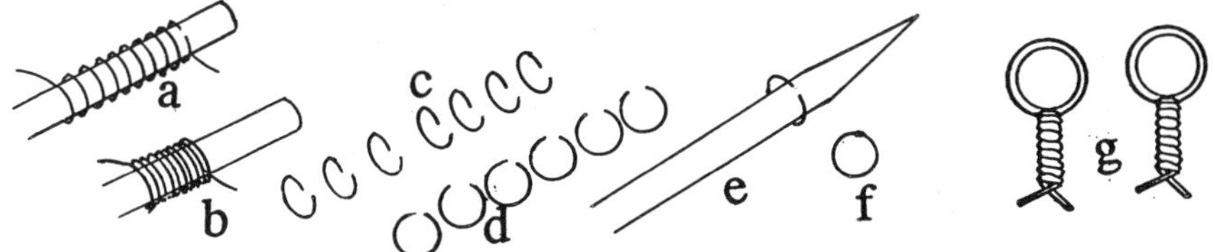

Fig. 44. Curtain rings and eyebolts for the curtain assembly. a-f. Making curtain rings: a-b. Wrapping wire coil around mandrel. c. Coil cut into separate volutions; individual volutions before straightening. d. Volutions straightened and ready for soldering. e. One volution (ring) on mandrel, ready for soldering. f. Finished ring (after soldering). g. Miniature twisted-wire eyebolts (0.25" long), to be used in guiding control thread for curtains.

Lead weights are absolutely essential to the smooth operation of the curtains and to their proper closure and opening. The ones shown in Fig. 45, d were cast in a dental-stone mold from a wooden pattern (a) turned on the lathe and cut longitudinally (b) with a jeweller's saw.

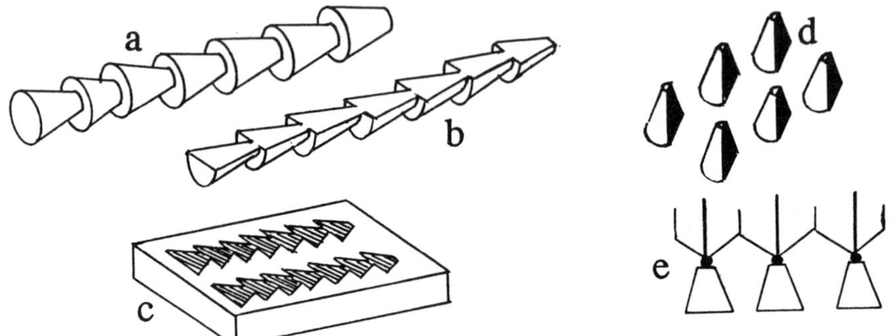

Fig. 45. Cast-lead weights for curtains. a. Hardwood pattern turned on lathe. b. Pattern produced by sawing the original in half longitudinally. c. Hard-plaster mould from which weights will be cast. d. Individual weights after casting and parting. e. Position for attaching weights to curtain folds.

The control thread for operating the curtains should be of sturdy material, mine being linen. By waxing it before use (beeswax), the thread will run more smoothly and freely through the eyebolts and its life should be increased commensurately. The thread should be run through the eyebolts set on the mounting board as shown in Fig. 46. It should then be attached with a

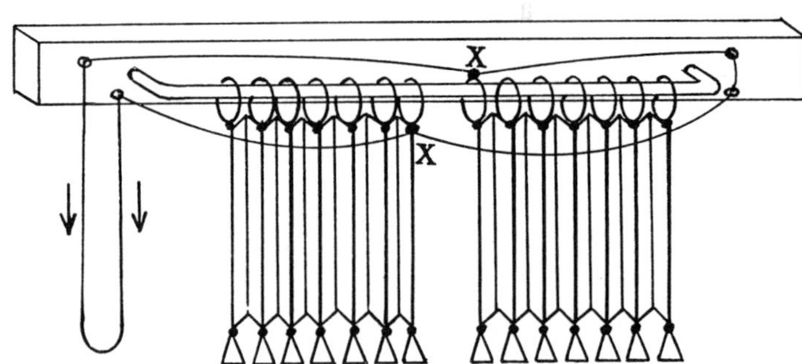

Fig. 46. The curtain assembly. The path of the control thread with eyebolts used as guides is shown, as are the points (x) at which the control thread is attached to the key curtain rings.

small drop of hot-melt glue applied at the points indicated by an *x* (Fig. 46), i.e., at the top of the leftmost ring of the righthand curtain and the bottom of the rightmost ring of the lefthand curtain.

After the entire apparatus has been assembled and adjusted until the curtains draw evenly and smoothly, it can be mounted by passing it through a large slot cut in the left side of the tube just behind the postscenium (frame) and inserting the right end through a slot cut in the right side of the tube. The tops of these two slots should be at a point sufficiently high up the side of the tube to insure that the bottom of the mounting strip is just above the path of the light. When the curtains have been straightened out and made to run freely by means of the control thread, and when the control thread has been brought to the outside for easy access and use by the operator, the mounting strip can be glued (carpenter's glue) to the top of the slots and clamped securely to dry. All is now in readiness for the first performance. If, however, as with mine, the curtains do not readily draw or close fully, the performance can be enlivened considerably by having a stagehand appear as the first character on the screen, who then uses a special hooked-wire tool to deal with the recalcitrant curtain—in full view of the audience, of course. With a little ingenuity or the part of the operator, this procedure can become the best part of the performance.

The range of possible subjects for one's shadow plays is vast indeed and provides a grand opportunity for an imaginative approach to miniature productions.

Should the construction of a draw-curtain system appear too difficult or too time consuming at the scale of the miniature theater, it is much easier to achieve efficient opening and closing of the stage through the use of a drop curtain instead (Fig. 47). A single curtain, large enough to cover the rectangular opening in the cardboard frame that produces the stage shadow on the screen replaces the draw curtain, a length of heavy wire solder inserted in the lower seam of the curtain provides the weight essential to efficient dropping and raising, and a length of 0.125" dowelling with wooden knobs at each end controls the curtain action. The control rod passes through holes cut in opposite sides of the body tube of the theater, these holes just large enough to insure easy curtain action. A slot cut across the top of the body tube enables one easily to attach the curtain to the control rod and can then be covered with a patch strip (piece of card stock glued to the tube) after installation. The postscenium should be mounted just behind the curtain, this best done before the curtain is installed.

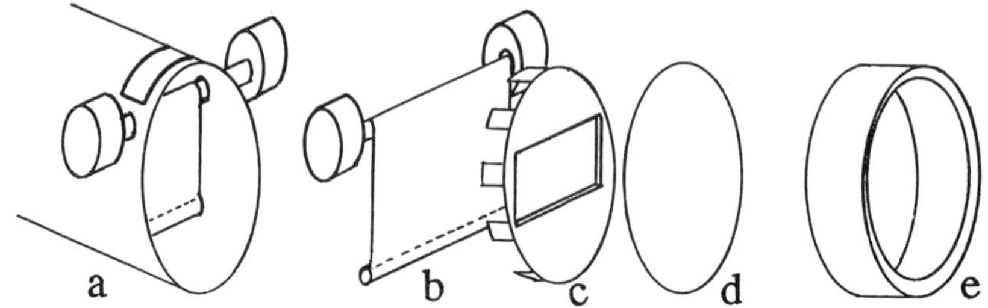

Fig. 47. A drop-curtain system for the miniature shadow theater. a. View of curtain and control rod installed and ready for use (viewing screen not yet in place, and stage frame (postscenium) not shown (it will be behind the curtain when installed). Note patch strip glued over the access slot after the curtain is attached to the curtain rod. b. The curtain assembly, weight at bottom provided by a length of wire solder. c. The postscenium. d. The viewing screen. e. The retaining ring for the viewing screen.

THE LILLIPUTIAN SHADOW THEATER

A shadow theater the size of ours
Is easy to carry or store,
And its grace and charm far o'ertowers
One that in size is much more.

The cast is no problem, they're all made of wire,
With bodies and clothes of plaster and paint.
Oh, yes, you can dress them in fancy attire
Without at the cost feeling faint.

As the miniature characters strut on the stage,
Propelled by your own fingers neat,
The folks will enjoy it no matter their age;
They'll find it a most welcome treat.

So just take the time to make yourself one;
It needn't be fancy or swell.
As soon as the first performance is done,
The crowds will be drawn by its spell.

PART SIX

A PICTORIAL GALLERY
OF SCULPTURAL MINIATURES

PAINTING AND SCULPTURE WED

Painting and sculpture, as everyone knows,
Are closely related in fact,
But to depict the relation in lilting prose
Requires a degree of tact.

Of course it depends on the medium involved
(material thermoplastic),
But once some simple problems you've solved,
You'll find the choices elastic.

The secret, indeed, lies in the tool
One uses to melt the stuff,
Solid as stone when cool
But fluid when heated enough.

If one is prepared to rely on color,
Fuse it all in with heat,
It's easy to change from one to the other,
Make paintings from sculpture so neat.

Of course it's all in miniature
(One's work is all done with a lens),
But as one developes skills that endure
It's easy to achieve one's ends.

So, pick up your tool and heat it up well,
Start melting the plastic waste.
I'll wager you'll soon make a picture so well
You'll switch to painting with grace.

17

BACKGROUND

The optimism inherent in thermoplastics technology and in its application to miniature sculpture, has, in my search for the numina of thermoplastics miniatures and for ways best to be true to the natural properties of the materials themselves, led to experimentation along various lines, including: (1) attempts at portraying the universal harmony in nature through realistic representation [starting with the production, from wax and clay originals, of cast polyester models of the lovely shells of microscopic marine protozoans (the foraminifera)]; (2) the production of copies of the works of sculptors representing various schools, from classicism to surrealism and on to impressionism and later; (3) automatic and semi-automatic sculpture (through the use of the pulled-adhesion process, for example); (4) the sculptural interpretation of blot painting; (5) silhouettes and cameos of sculptural pieces viewed either directly or as cast on a miniature screen; (6) sculptural puns; (7) purely fanciful subjects; and (8) jewelry. These areas of experimentation will be considered in sequence, and examples of some of the products of each will be given in the plates which follow.

The realistic depiction of nature

A nice illustration of the existence of the type of "universal harmony" that motivated Brancusi when he created the elegant simplicity of his *Bird in Space*, and one which seems also to emphasize (as did he and Barbara Hepworth) the importance of gaining an understanding of and developing a respect for the nature of one's materials, is the ease with which one can, through the development of anlagen made with the pulled-adhesion technique, produce successful imitations of the stalactitic and stalagmitic growth of the cave deposits of percolating ground waters, a delightful example, too, of being true to the nature of one's materials. In this case, the resemblance between nature (the cave deposits themselves) and the thermoplastic forms is a superficial—but no less striking—one, but, at a miniature scale, one is conscious principally of the overall resemblance in form and not of the relatively minor and insignificant differences in surface texture that betray the fundamental differences in origin between the works of nature and those

155

of man. Yet, even when one considers the basic differences between the two processes, one also recognizes (as Brancusi did) the universal harmony that binds the two: the compliance with the physical laws that underly the growth process, the operation of physical forces "on matter animated by internal energy", whether this energy be that of crystallization in percolating ground waters or the molecular forces of plastic flow in molten polyethylene. In this case, similarities arise from the response of both to the various natural forces: gravity, surface tension, friction, crystallization, as well as to hydrostatic, electrical, molecular, atomic, and ionic forces. The craftsperson or artist supplies special forces in producing the plastic imitations of nature, however, among them the initial compression that causes the molten mass to adhere to the wooden plates between which it is sandwiched, and the tensional force that pulls the mass apart. Both of these forces are mirrored in the finished piece after it cools, and they contribute in no small measure to the overall resemblance between the natural and the artificial products. Vestiges of the initial compressive stage are apparent in the area of contact between the plastic and the wooden surface to which it adheres, while the tensive force finds expression in the extenuated intervening form between and in the lines and folds of stress developed throughout in the heat-softened mass. The form that results from the combination of the two compression areas of attachment (above and below) with the intervening tensile area as the mass of heat-softened plastic is pulled in opposite directions is analogous with the overall form of the stalactite/stalagmite system in which the repose of the latter and the dynamic form of the former basically reflect their individual responses to gravity.

While it is easy to produce convincing stalactite-and-stalagmite-rich miniature grottoes in imitation of real caves or the artificial ones which (as was pointed out by Vasari in his book on the techniques of the artist, architect and sculptor) so delighted the patrons of renaissance sculptors, any hope I might have of reproducing in miniature anything in satisfying mimicry of the highly ornate stalactitic clusters of corbellated squinches found in such jewels of Islamic architecture as the arcades, the Lion Court and the Baraxa Tower of the Alhambra or the great *iwan* in the courtyard of the Masjid-i Jami' in Isfahan is infinitely more remote than my simple desire—still discouragingly frustrated by my lack of technical skill—to copy in miniature some of the sculptural gems found on smaller Wedgewood pieces.

The two basic geological processes of building up and wearing down rocks, minerals, and strata, processes by which, on the one hand, sedimentary products are built up (such as concretionary growth and stalactite/ stalagmite formation, for example) and, on the other, by which rocks, minerals, and strata are broken down or worn away (the processes of weathering and erosion, for example) are easily analogized with the modelling and carving procedures respectively of the sculptor. To a geologist the analogy is a striking one, particularly when viewing such sculptural pieces as Medardo Rosso's stalagmitic *Conversazione in Giardino,* Jean Arp's *Human Concretions,* Barbara Hepworth's pierced forms, Piotr Kowalski's *Calotte 4,* and Alberto Viani's *Torso.*

(As an aside, though still fascinatingly relevant, it is quite intriguing to note in passing that this analogy is actually reflected in the etymological relation between the words *sculpture* and, in its geological or topographical sense, *shelf.* To a geologist, a shelf often is a structure formed by eroding rocks away to produce a flat expanse, which may range in size from a small section of wave-cut bench along some California seashore to the massive coastal plain that covers so much of the southeastern United States. Here Nature was the sculptress, the process itself the precise analogue of the glyptic artist of classic Greece whacking away at a piece of marble to give birth to a *Nike of Samothrace.*

Still on a mundane—albeit microscopic—level, we might seek our inspiration from such natural phenomena as the minute amorphous and pseudomorphic masses of the mineral glauconite (an iron photassium silicate) and the microscopic faecal pellets so often replicated in it which, for the microscopist, seem to find large-scale expression in such sculptural works as Olga Jančić's *Double Form,* Jacques Lipchitz's *Benediction I* and Henri Matisse's *Tiaré With Necklace.*

The analogy between natural processes and the procedures of the sculptor, however, is not confined to geological areas, but extends to biological ones as well, as indicated here:

Geological	Biological	Sculptural
concretionary growth	*anabolism*	*modelling*
stalactitic/stalagmitic growth	*growth*	*modelling*
erosional processes	*catabolism*	*carving, filing, etc.*

Using conventional sculptural techniques (i.e., carving and modelling, although here achieved with unconventional tools and procedures: the electric-loop tool, etc.) and turning from the geological to the biological world, one finds that certain types of natural organic forms are more easily produced than others. Some of the *protean* forms one encounters in many of the lower organisms—particularly among the Protozoa, not surprisingly, to judge from the name applied to the group—are easily produced in these materials. Indeed my own experience in sculpture began with attempts at reproducing as teaching aids the microscopic shells of marine protozoans (particularly the foraminifera) at a size easily seen without a microscope. The original modelling for these was done in wax or clay, silicone-rubber molds were made from these wax patterns, and centrifugal castings in polyester resin were finally produced with the molds.

Ernst Haeckel's work (1909) here is, indeed, a massive source of designs from nature—many of them overwhelmingly complex, however—from which an ambitious miniaturist migh seek inspiration. Among these are some of the forms (of Foraminifera, for example) that appear in Plate 14 as completed pieces in my collection.

As one moves up through the animal kingdom (as Haeckel does in his collection of drawings from nature), one passes through many groups having members that are relatively easy to duplicate in miniature sculpture, and well up in the tree of life one encounters forms that are particularly easy to reproduce in thermoplastics sculpture. When the entire body of the higher mammals, including man, is to be represented, twisted-wire armatures are helpful, but busts are relatively easily produced from a solid mass, employing a combination of carving and modelling techniques.

Imitating or copying existing sculptural works

As with conventional sculpture, excellent experience may be acquired by attempting to make miniature copies of well-known sculptural pieces. In this way one learns which techniques and tools are appropriate in achieving convincing results in a wide variety of sculptural styles and effects.

If one is working from pictures of sculptural pieces rather than from the actual piece, the simple copy holder illustrated in Fig. 4, g can be most helpful, because with it the picture is held convenient to the work area, ready for constant reference as one's work progresses.

Some sculptural works are much more easily imitated or copied than are others. Several of Jean Arp's simpler pieces, such as his *Star, Human Concretion, Landmark,* and *Ptolemy I* posed no real difficulty, nor did Brancusi's *Mademoiselle Pogany, Bird in Space,* and *Bird.* Interestingly, such pieces as Rodin's *Burghers of Calais*, Rosso's *Conversationi in Giardino* and his *Bookmaker*, and Canova's *Meekness* also were relatively easy to miniaturize; possibly the fact of their being roughly carved and in a general manner resembling the form of a stalagmite—a massive basal section tapering and narrowing upward—simplifies the miniaturist's task. The same ease with which manufacturers of imitation (plastic) marble today impart the various markings found in natural marble as the consequence of impurities and mineral segregations was easy to achieve in copying Hajdu's *Delphine*, the markings here added simply by melting extruded threads of darker plastic into the white body of the piece and, with a needle, shaping and moving them into position in a locally-melted area to create an appearance not unlike, or at least somewhat suggestive of, garnet segregations or pegmatite dykes in granite.

Among other work that was easily copied in miniature were several of Barbara Hepworth's and Henry Moore's pieces, Dalí's *Architectonic Angelus of Millet*, Matisse's *Tiaré With Necklace*, Cascella's *The White Bird*, and Martins' *Rituel du Rhythme*, although the last was more easily formed on a wire armature than without it. To reproduce the *Nike of Samothrace* in miniature was an irresistible challenge, to achieve acceptable results actually easier than anticipated.

A miniature copy of Boccioni's *The Mother*, using the electric loop tool, a cold needle tool, and a flat-bladed tool (cold), was reasonably acceptable. Indeed a great deal can be done with this basic technique and these few tools, but one needs some special tools and refined techniques to shape such intricate surface contours, flats, and hemispheres as those found on a piece like Boccioni's *Unique Forms of Continuity in Space*. Replicating the many armor plates required an approach I had not yet attempted. Although the individual plates vary greatly in size and overall shape, most of them are esssentially curved surfaces and look as though they had been shaped over mandrels by which complex combinations of convex and concave surfaces have been developed. In sheet metal, this can be done with ease, using shaped mandrels or stakes, but to accomplish the same thing in polyethylene is more difficult unless one can develope techniques particularly appropriate to the task. My initial attempts at producing such plates with the electric-loop tool and a needle were promising but tedious. Obviously it could be done more efficiently in other ways. I then made a two-piece mold in dental plaster from wax patterns and used the molds to press out plates from masses of heat-softened plastic, a promising but time-consuming procedure.

Even more discouraging have been my efforts at producing acceptable miniature copies of some of the intricate cameo-like carvings found on the smaller pieces of Wedgewood pottery, my best efforts still terribly crude in contrast. At room temperature it is, for me at least, much more difficult to carve the cold plastic with a cold tool and achieve the well-finished surface detail of the Wedgewood miniatures than it is to do so in the proper sculptor's wax. Instead of carving away minute quantities of the plastic as one would with wax and then remelting the worked area carefully with the electric-loop tool to produce the desired smoothness, I find it easier to remove the bulk of material in its molten or heat-softened state (without deforming the remainder through excessive heat and melting) and then proceed to the final shaping after the plastic cools. This is the technical frontier that still defies me and prevents me from achieving the standard I am able to reach in making wax patterns and then centrifugally casting polyester resins either directly from silicone-rubber molds of these patterns or through the lost-wax procedure. I dare say, however, that a more skilled miniaturist should be able to develope an adequate technique and achieve acceptable results with such a demanding copyist's assignment.

Among the challenges for the copyist working with thermoplastic miniatures would be sculpture gardens with a collection of pieces ranging from classical Greece to such moderns as Hepworth and Moore. One might, instead, prefer to reproduce Japanese netsuke or, perhaps, cameos from classical times to the present. The range of possibilities is limitless.

Automatic and semi-automatic sculpture

The form taken by the earlier stages in creating miniature sculptural pieces through the use of the pulled-adhesion procedure described on pp. 57-61 depends to a certain degree on factors and forces not easily harnessed, so to this extent such pieces are automatic or semi-automatic (these two terms used not in the conventional sculptural sense but simply to indicate that some phases of the operation are not under the conscious control of the operator), depending upon how much one modifies the anlagen developed from the urschleim through pulling. Some control can be exercised over certain of the factors involved in producing the urschleim itself (temperature, volume of plastic, type of substratum used, etc.) as well as in producing the initial form of the anlagen, and after the form resulting from pulling and cooling has been produced the sculptor's imagination takes full charge.

The sculptural interpretation of blot painting

A large number of forms developed from the blots produced as described on pp. 121-124 and illustrated in Fig. 35 are shown in Plates 1-5. It was a relatively simple matter to turn each of the blots into a sculptural piece, the various technical challenges they presented a wholesome learning experience.

Silhouettes and Cameos

It is quite challenging to produce sculptural miniatures from silhouettes: from those cast as shadows in the shadow-viewer shown in Fig. 42, a, from those one can cut oneself from black paper, using as a subject any person or object that might be available, or from pre-prepared ones from books or clip-art sources. To render such pieces in cameo form can also be a delightful challenge. If one can survive the oxymoron, *miniature* medallions are also fair game for the enthusiastic thermoplastics miniaturist.

Sculptural puns

The challenge of punning with sculptural pieces is tempting, as suggested by those shown in Plate 13. The *Holey C* (Holy See) shown in Plate 13, a, is one such example, as is *The Star-Spangled Banana* seen in *k* of that plate, both of these slightly moronic but fun to make and reasonably good laugh-or-groan getters when shown to tolerant friends. (The banana skin was made of yellow plastic from the opening strip of a tamper-proof, vacuum-sealed jar of peanuts, to contrast realistically with the edible part of the banana. Stars were simply paper cutouts attached with hot-melt glue.)

More effective still are kinetic puns, such as the *Vibraphone* shown in 13, b. The case of the phone was built up from flat scrap, the handpiece from extruded rod stock, the dial punched from flat stock and drilled (10 holes) before being attached with a small drop of hot-melt glue. The entire phone was attached to a thin piece of spring wire inserted into the end of a short length of applicator stick. By flicking the free end of the stick, the phone is made to vibrate energetically, whence the pun on the xylophone-like musical instrument. A play could also be made on *Vibraharp*, using either a harmonica or an Aeolian harp as the counterpart of the telephone. Some of the numerous devices described in Arnold's *Musical Punstruments* should be easy to convert into sculptural miniatures, *The Clef Hanger, O Holey*

Knight, Awl Through the Knight, Good Knight's Sweetheart, and *Jello, Dalí,* for example, but some of the more complicated ones, particularly such kinetic ones as *The Compact Disk Jockey, The Wetting March,* the various ones based on the word *Band,* and *The Orange-Juice Harp,* may require some simplification or other modifications before successfully being reduced to miniatures. Simple sculptural puns that survive reduction to silhouette form may effectively be presented with the shadow-viewer, the animation so easy through its use being a great advantage. If one's puns are musical (and no matter how they are displayed), the performance is enlivened by producing musical clues to their identity (the audience attempting to guess the title or the pun involved), the music supplied by humming, whistling, singing, or by playing a miniature harmonica (for which most of the musical punstruments described in the book were designed), a chromatic harmonica, or any instrument of one's choice and appropriate to the occasion.

Anyone intrigued by puns should become acquainted with Redfern's (1984) scholarly study of this fascinating art of language mutilation and mental gymnastics. His "second thoughts" (1996) extend the adventure.

Fanciful miniatures

Thermoplastics, worked with tools and techniques described in this book, lend themselves to the development and portrayal of fanciful ideas and designs, as suggested by some of the fanciful elements to be seen in various of the illustrations in the accompanying plates. A more imaginative and artistic worker could undoubtedly produce pieces of real artistic merit, unlike those of my amateurish efforts, but my intention here has merely been to indicate something of the scope for creative development in this recycling art form.

Miniature Sculptural Pieces as Jewelry

The craftsperson or artist interested in designing and making jewelry may well find challenges in the use of miniature sculptural pieces as jewelry or ornaments of various types. While the thermoplastic materials are, admittedly, less sturdy than most metal or stones normally found in jewelry, they can, if worked in suitable design and with proper respect for their delicacy, prove quite effective and serviceable and are most certain to evoke comment and admiration. They make fine conversation pieces.

PLATES

Many of the miniature sculptural pieces illustrated here are bozzettos, relatively rough sketches, requiring additional attention. Each of the illustrations has its sculptural counterpart, however crude or incomplete, in my collection. The pieces and the illustrations of them are intended principally to indicate something of the range of form and subject one can attempt in this medium and at diminutive size, in the hope that others more imaginative and dexterous than I will pursue the craft or art to its proper conclusion and mastery or derive at least as much pleasure from striving to do so as I have.

PLATE 1

Note for plates 1-5: Each of the drawings on these plates consists of two parts: an ink blot and a line drawing, the latter an interpretation of the former, with an arrow between them to indicate the relationship. The longest dimension is given for each completed sculptural piece derived from the interpretation and here illustrated.

a. *The Capture*   1.2"
b. *In Playful Mood*   0.7"
c. *The Philosopher*   0.8"
d. *The Parachute Bird*   0.7"   The parachute was made by free-
   blowing a low-density-polyethylene blank cut from sheet stock.
e. *The Escapee From Jurassic Park*   1"
f. *Gentle Persuasion*   0.7"   Both man and horse were modelled on
   twisted-wire armatures.

PLATE 1

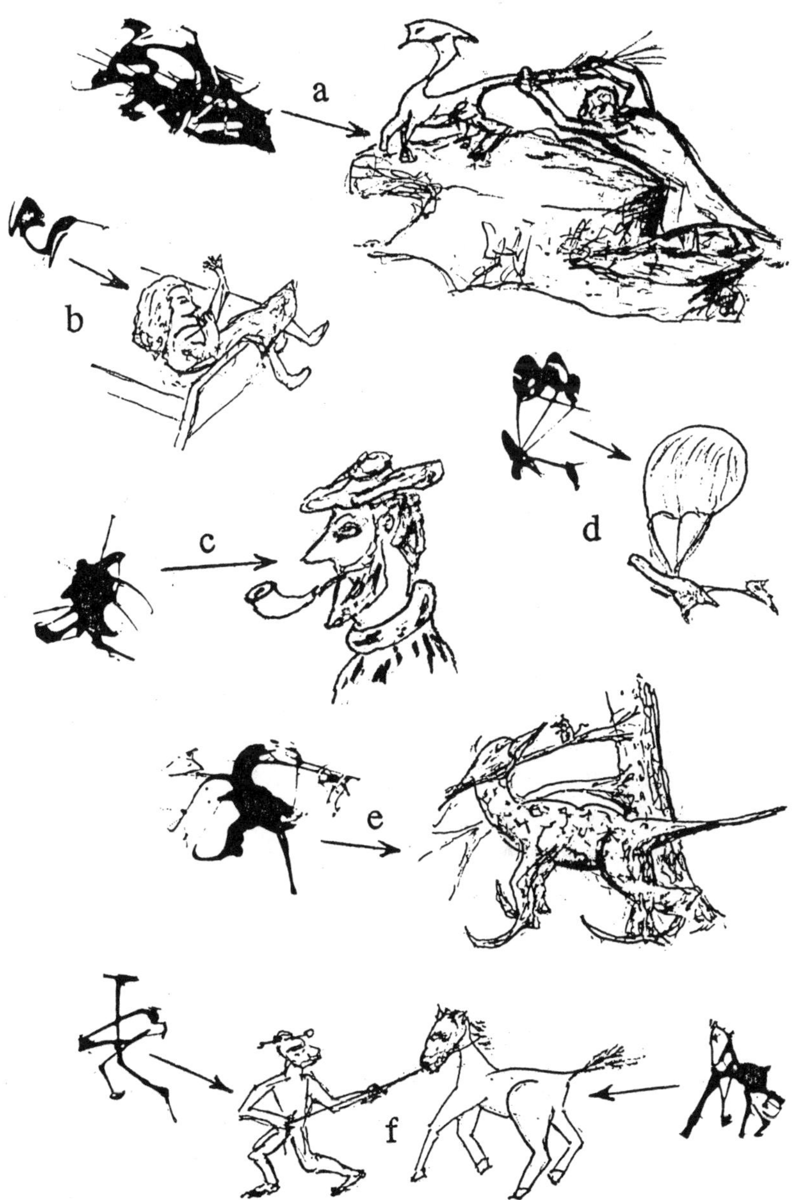

PLATE 2

a. *En Retraite* 0.7"

b. *Consolation* 0.9"

c. *Nefertiti's Cousin In The Khamsin* 0.7"

d. *La Coiffure Ineffable* 0.6"

e. *African Queen Waiting For Humphrey Gobart* 0.7 *Humphrey Gobart* is the name lovingly given the campus shuttle bus at U.C., Berkeley.

f. *La Confrontation Gallinacé* 1.1 "

g. *It's Mine!* 0.7

h. *Bird In Flight* 0.2"

i. *Pride* 1.2"

j. *L'Oiseau Prophète* 1"

k. *Voilà, mon chat!* 1"

l. *L'Enfant Terrible* 1"

m. *The Rampant Stallion* 1.2"

PLATE 2

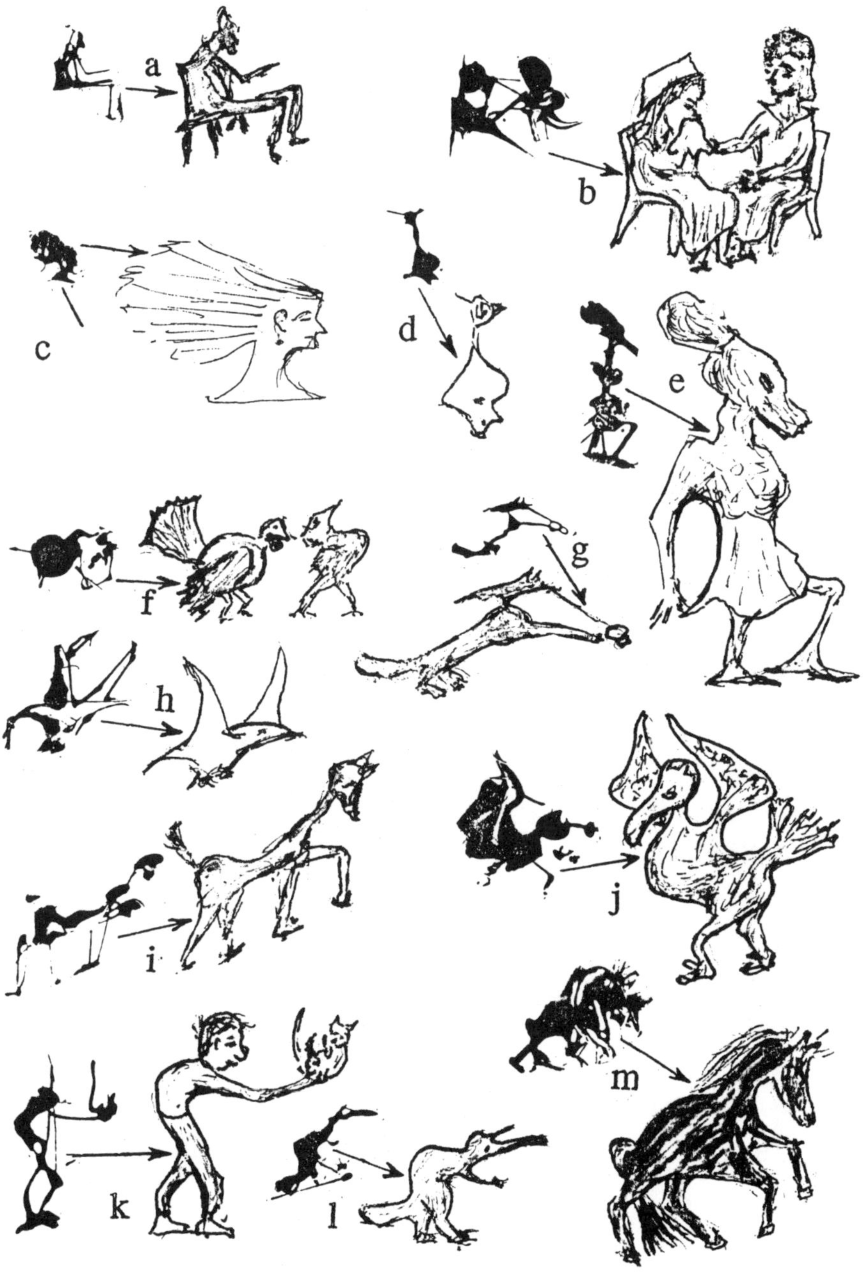

PLATE 3

a. *Christine* 0.6"
b. *Puss in Boots* 0.7"
c. *Tethered I* 1"
d. *Witch's Dance* 0.7"
e. *Vengeance Is Mine* 1.5"
f. *The Kite Flyer* 1.1"
g. *Estelle* 0.8"
h. *Insouciance* 0.7"
i. *The Peripatetic Monk and His Ardent Follower* 0.8"
j. *Tethered II* 0.7"

PLATE 3

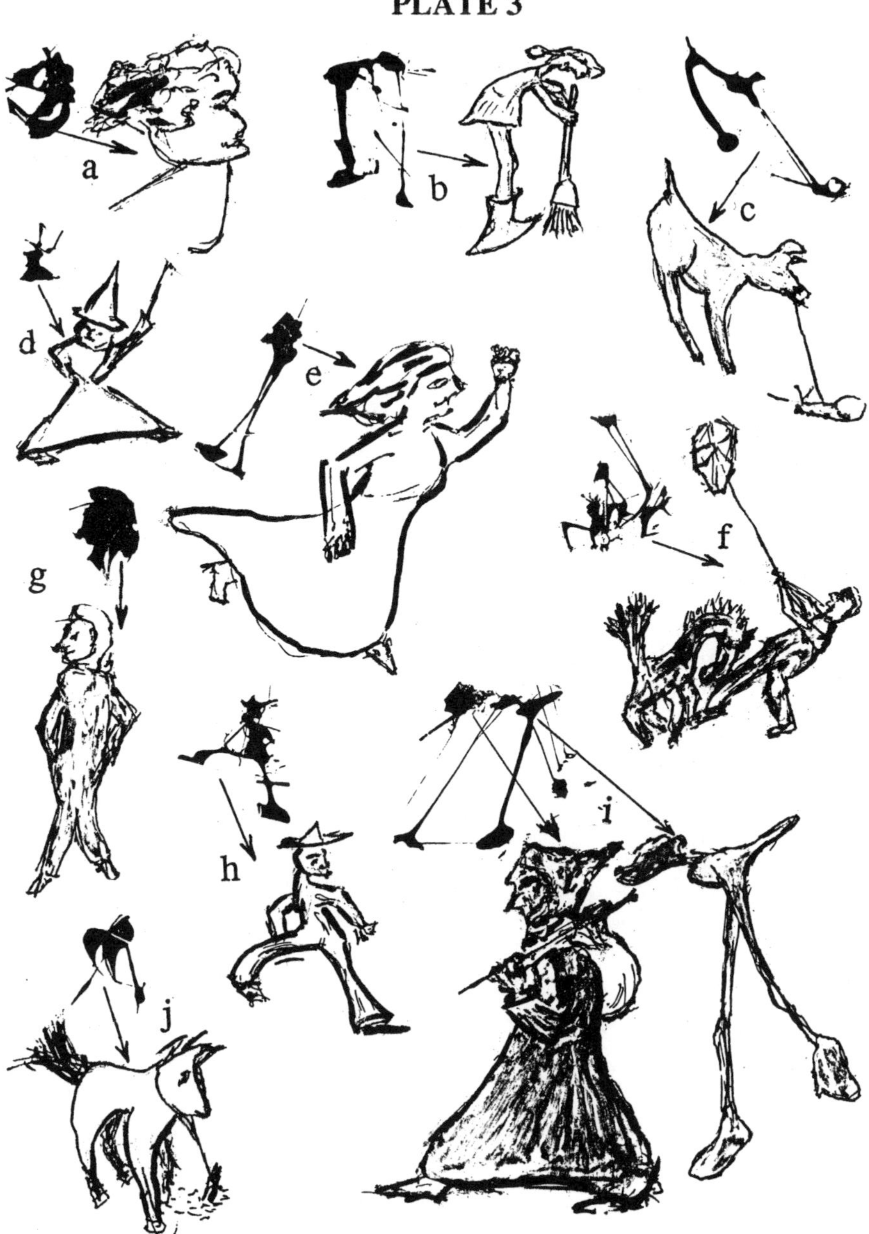

PLATE 4

a. *The Observer* 0.5"
b. *Reluctance* 0.8"
c. *Christmas Scene*   0.1"
d. *The Skier* 0.6"
e. *Le Secret* 0.8"
f. *Night on Bald Mountain* 0.5"
g. *The Strawberry Roan* 0.5"
h. *The Compleat Angler* 0.8"
i. *The Cavorting Witch* 1.1"
j. *The Artist I* 0.8"

PLATE 4

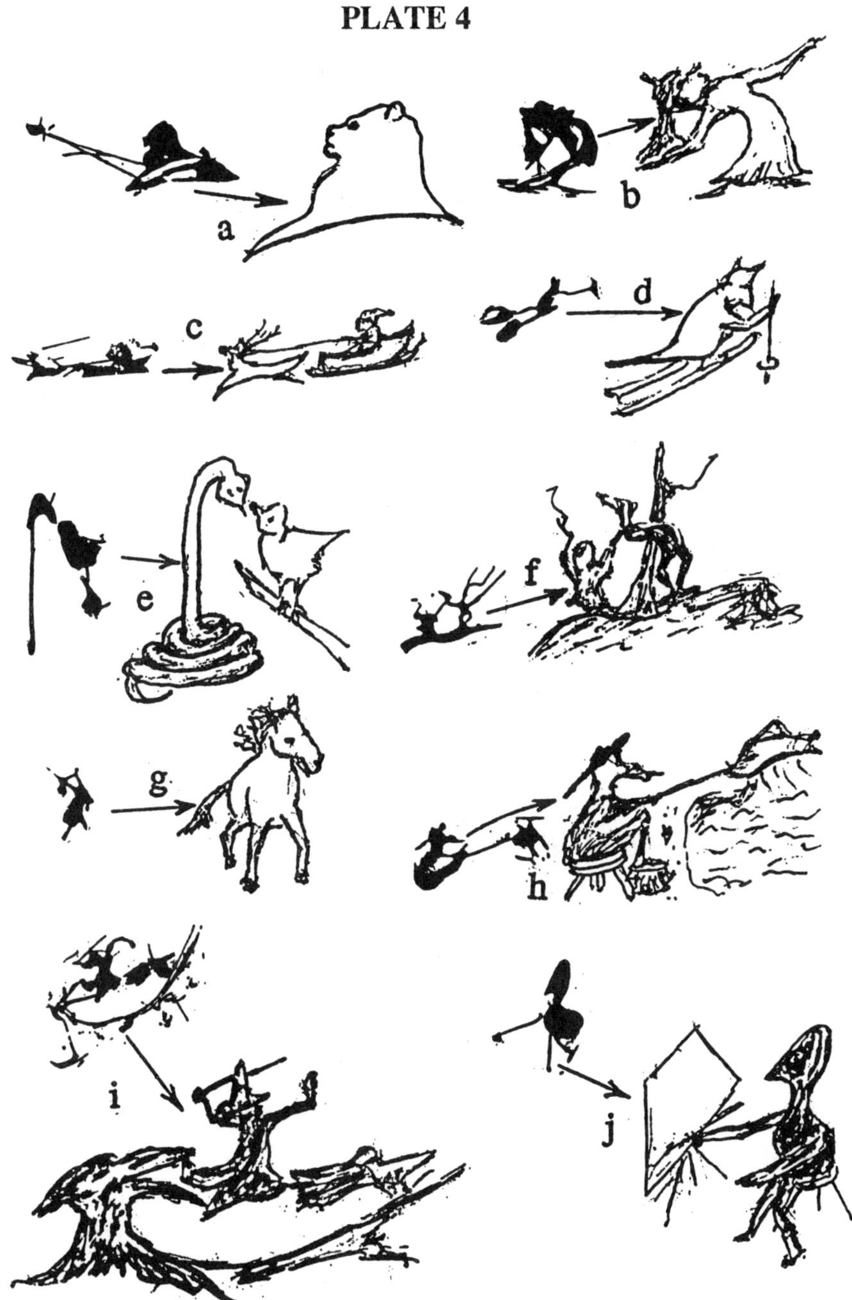

172

PLATE 5

a. *L'Artiste II*   0.6"

b.*The Unsociable Climber* 0.8" (The climber is mounting a thin steel wire.)

c. *The Rollicsome Twosome* 0.7"

d. *The Religious Schism*   0.8"

e. *Nude in Repose*   1"

f. *Bambi's Aunt Sophie* 0.8"

g. *Pals* 0.7"

h. *Surprise* 0.5"

i. *Adumbration of Speed*   1"

j. *Lorenzo the Magnificent's Plumber* 0.7"

k. *Show Me The Way To Jurassic Park*   1.1"

PLATE 5

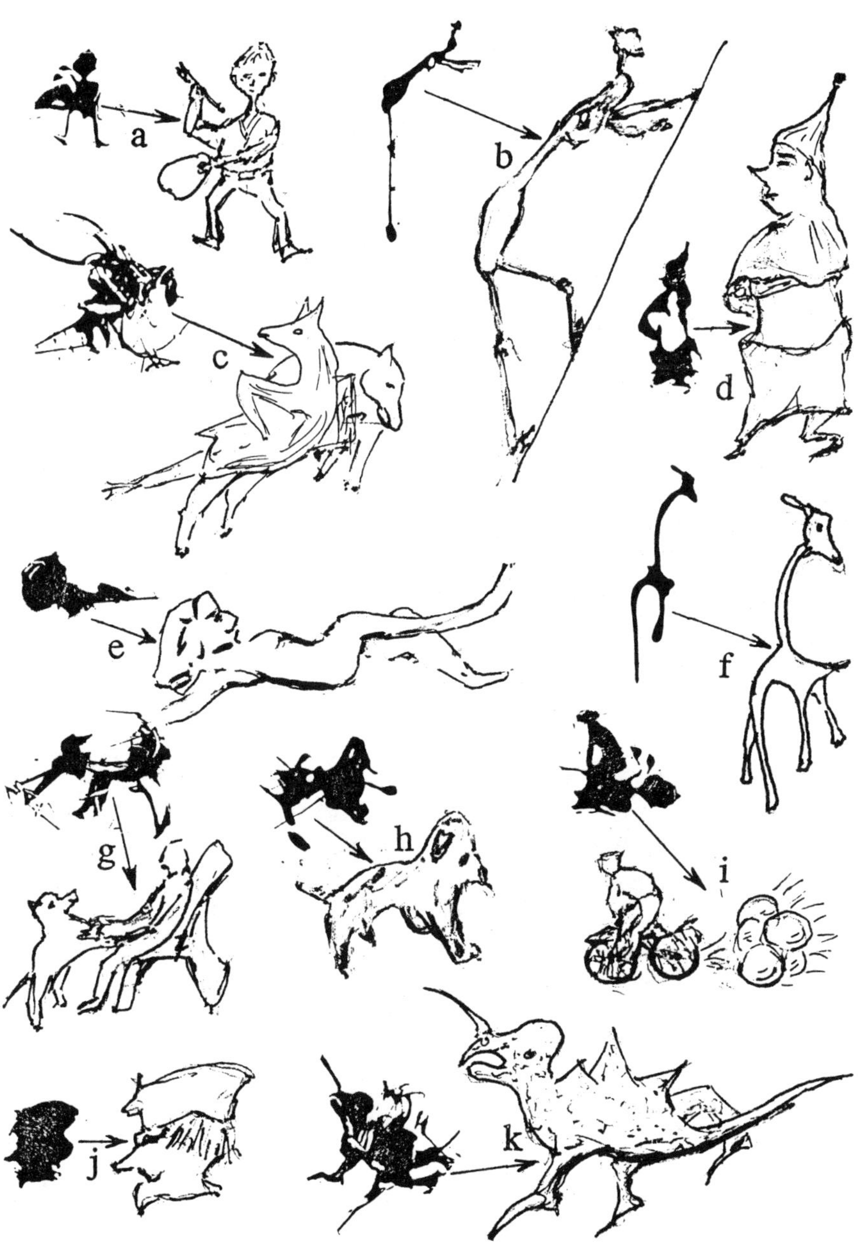

174

PLATE 6

a. *Swan Song* 0.8"

b. *Winter Frolic* 1.2"

c. *The Bi-Belled Tubaphone* 0.7"

d. *The Church Exhalted* 2.1" This, one of my earlier pieces (April 28, 1978), was carved from a polystyrene block, using the electric-loop tool (the church, house, and bridge, however, carved from wood), but the difficulty of obtaining the desired detail and finish led to experimentation with polyethylene, a material which proved highly promising and led into the study represented by the material presented in this book. Note, too, that the size of this piece is two or three times that of many of the polyethylene pieces.

e. *Invitation To The Dance* 1.2"

f. *Don Quixote* 0.8" The Don's spear is a piece of brass rod stock. Patterned after a small wooden novelty purchased in Spain in 1963.

g. *Space Dance* 1"

h. *The Celestial Harp-Repair Shop* 1.1" The harps are built up on twisted-wire armatures, the strings are of fine wire.

i. *The Crossbye Prime of Slakes*   1" The replicated features were made by pressing hardware cloth into a mass of molten high-density polyethylene, an example of divisional (or subdivisional) sculpture.

j. *The Recumbent Debate* 1.1"

k. *Space Village* 1.1" The individual buildings were formed by blowing a heat-softened blank of low-density polyethylene into separate molds. See Fig. 21 (p. 80) for further applications of the technique.

PLATE 6

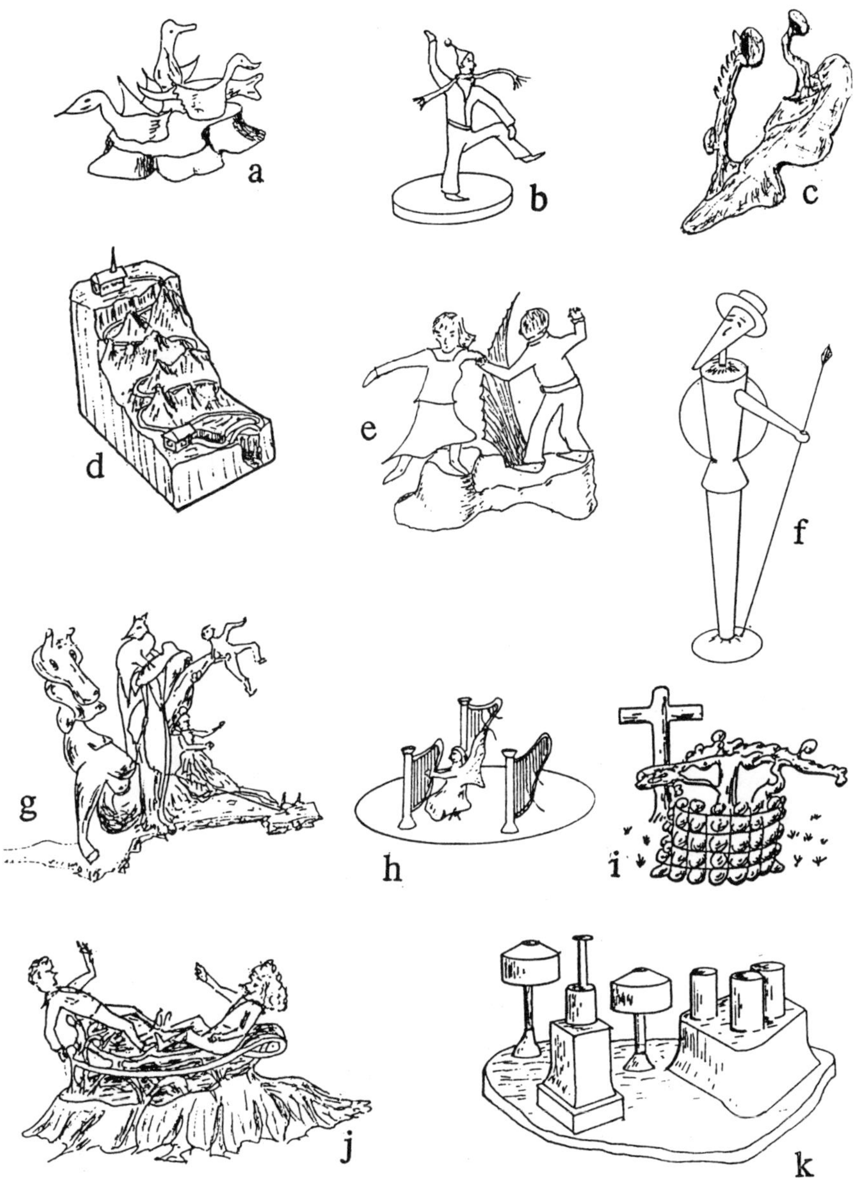

176

PLATE 7

a. *The Kite Flyer* 1" (Two men on a green-eyed dragophant hunt told the young lad who wished to accompany them to "go fly a kite", which he is now happily doing.)

b. *Morning Becomes Elektra* 1"   (The setting for this was made through the pulled adhesion process.)

c. *The Umpire* 0.7"

d. *The Slow Learner* 1"

e. *The Fantail Dance*r 1.5"

f. *Terpsichore's Children* 1.75"

g. *The Bird Watcher* (or *The Watcherbird*) 0.6"

h. *Ecstasy* 1.7"

i. *The Slings And Arrows Of Outrageous Fortune* 1.2"

j. *Diana of the Hunt* 1.7"

k. *A Nameless Fear* 1.5"

l. *The Bully* 1.6"

PLATE 7

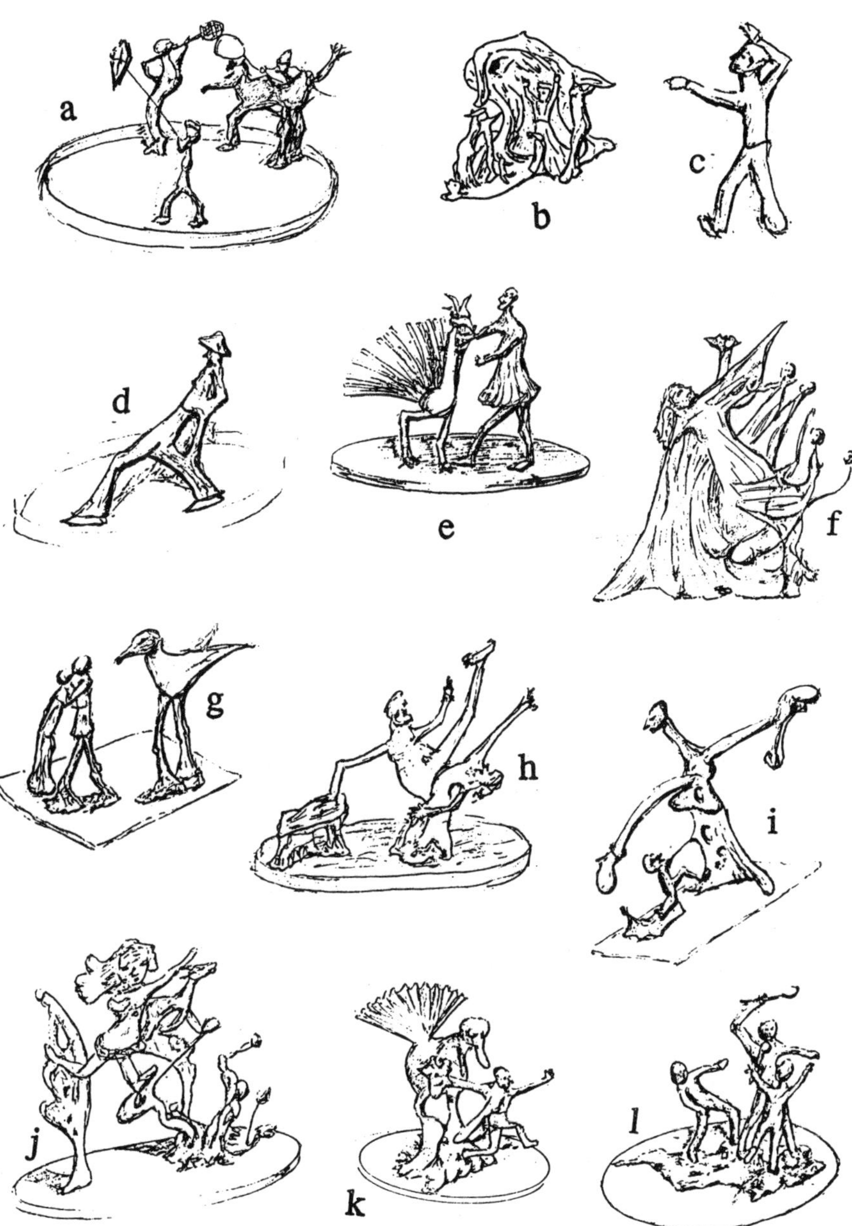

PLATE 8

a. *The Scarf Dance* 0.7"
b. *The Hitchhiker* 1.5"
c. *The Mermaid Sejant* 0.7"
d. *Sunrise on Easter Island* 1.6"
e. *The Acrobats* 1"
f. *Post-Conflagration* 1.2"
g. *Desert Song* 1"
h. *Enivrement* 0.8"
i. *A Curious Tale* 1.2"
j. *Le Parasite* 1.2"
k. *Impolitesse* 1.2"
l. *Aménité décapité* 1.2"

PLATE 8

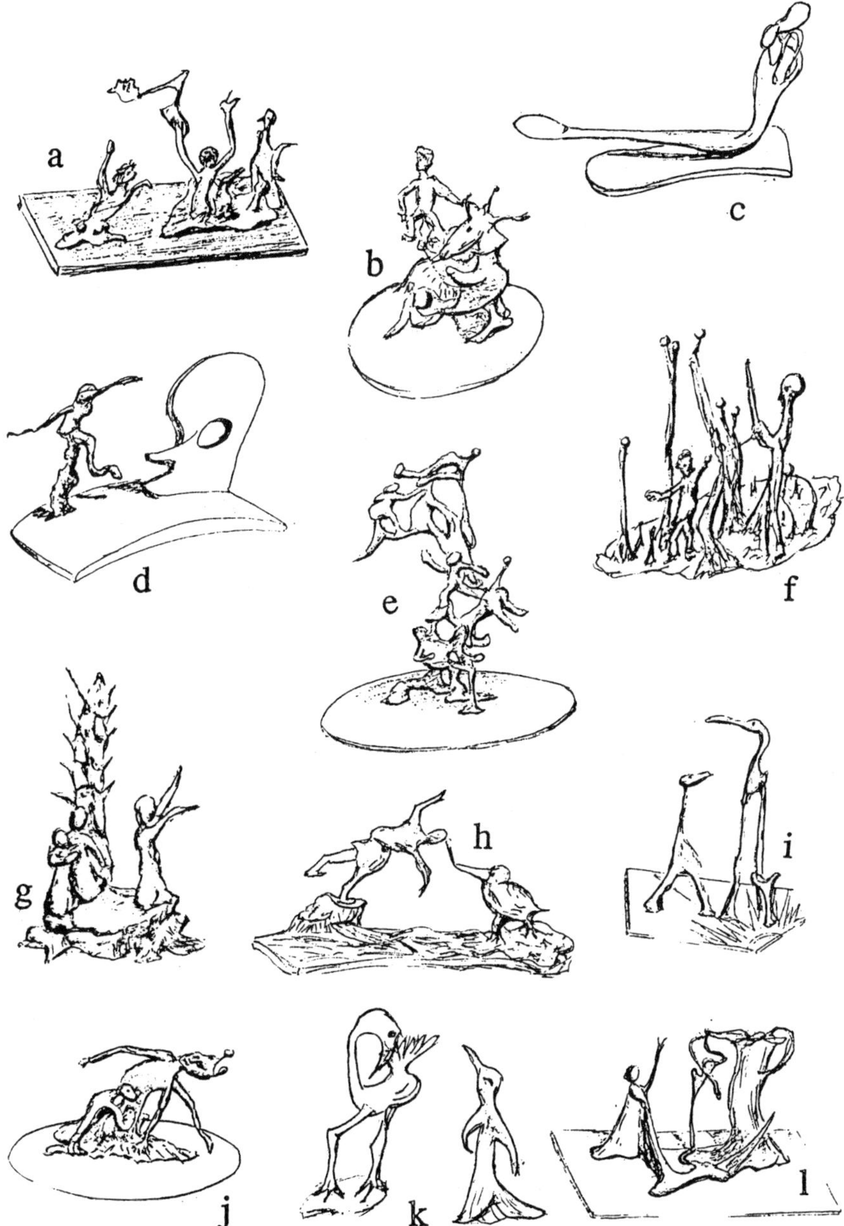

PLATE 9

a. *Jurassic Park Revisited* 1"
b. *Aspiration* 1.2"
c. *Desolation* 0.7"
d. *Persuasion* 0.8"
e. *Yggdrasill and the Corbelled Squinch* 1"
f. *The Probe* 0.7"
g. *Hommage à Hepworth* 1.5"
h. *Idyll* 0.7"
i. *Transmorgrification* 0.8"
j. *Espieglerie* 1.7"
k. *The Circus* 1.5"
l. *Le Béjaune* 0.6"
m. *Chinoiserie* 1.2"

PLATE 9

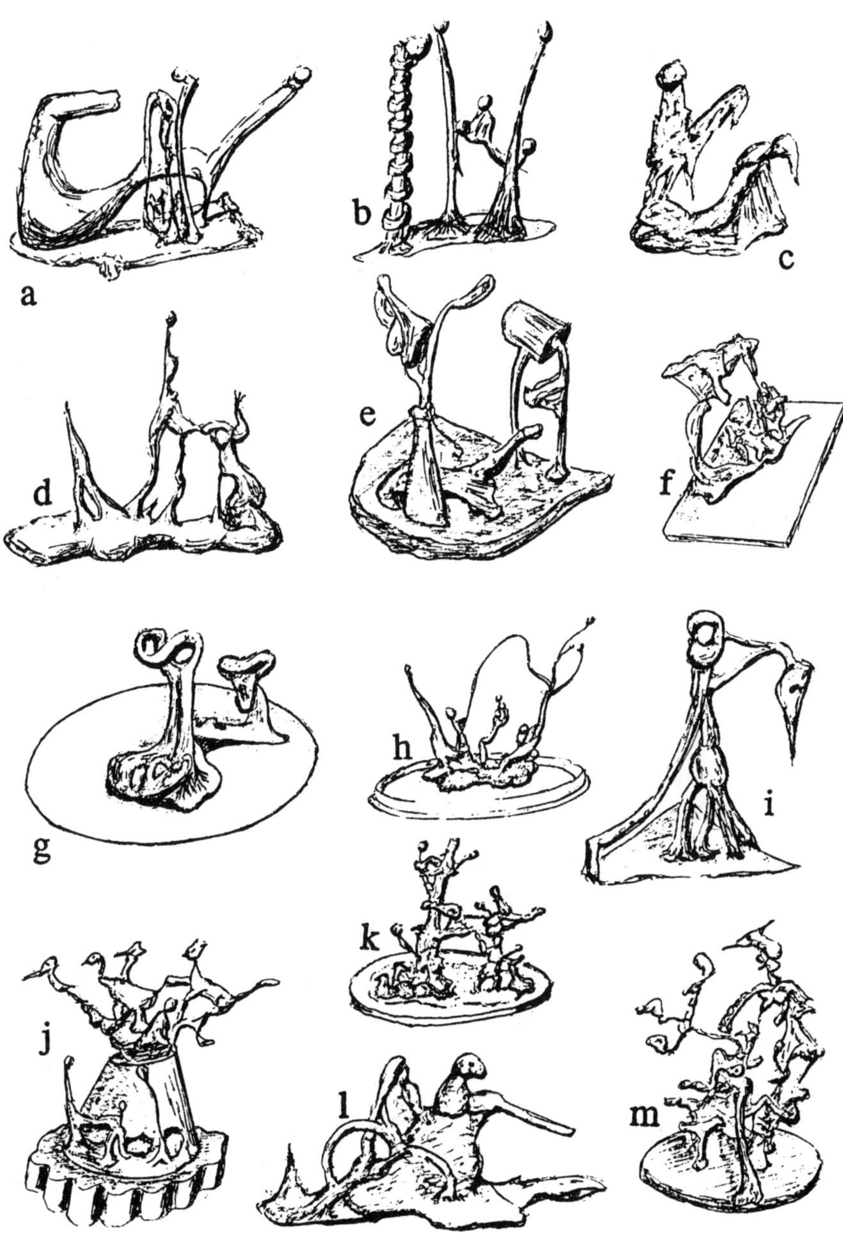

PLATE 10

a. *Le Pedant Chinois*   0.7"
b. *Whoa!* 0.7"
c. *The Race* 1.5"
d. *The Dancing Class* 1"
e. *She Did It!* 0.7"
f. *Crêche*   1.2"
g. *L'Annonciation* 0.6"
h. *Chloe* 1.2"
i. *Arrivals and Departures* 0.7"

PLATE 10

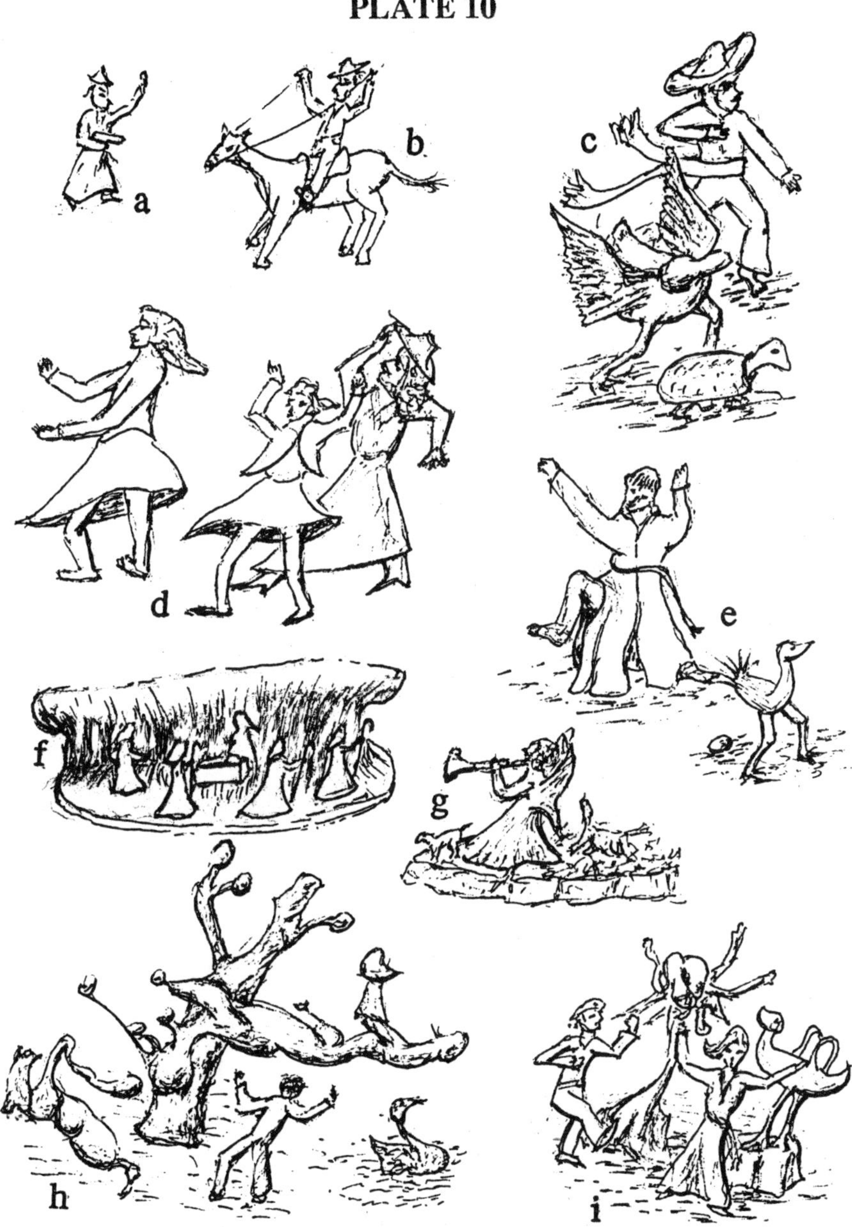

PLATE 11

a. *L'Innocence* 0.8"
b. *Le Faux Pas* 0.7"
c. *La Préceptrice* 0.8"
d. *The Outsider* 0.7"
e. *The Artful Dodgers* 0.6"
f. *La Jeune Fille* 0.5"
g. *La Beauté* 0.7"
h. *Berceuse* 1.6"
i. *Disbelief* 0.5"
j. *Apprehension* 0.8"
k. *Au Travail* 0.7"
l. *Prise de Bec* 0.6"

PLATE 11

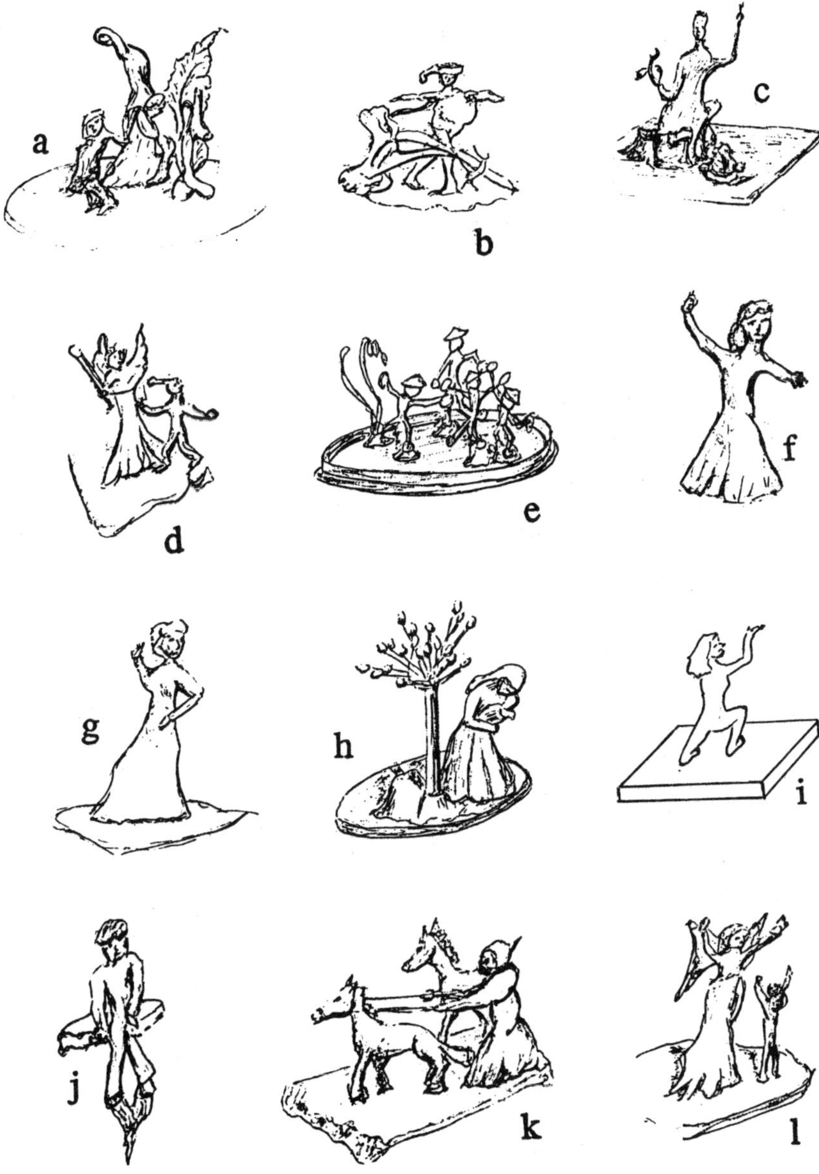

PLATE 12

a. *Remote Control* 0.8"
b. *Pas De Deux* 0.7"
c. *The Washboard Band* 1.2"
d. *From Aix to Ghent* 2"
e. *The Tuba Ischium* 0.7"
f. *A Proboscidean Interlude* 1.2"
g. *The Supralapsarian Selectorate* 1"
h. *Les Amis* 1.1"
i. *The Helix of Life (DNA)* 2"
j. *Houri, Hooray!* 0.8"
k. *The Achievers* 1.5"
l. *The Ferris Wheel* 1.7"
m. *Consternation* 1"
n. *Syllabub in Hades* 0.7"
o. *Shades of The Siegfried Washline* 2"   The base is an inverted walnut-shell half, the foraminifer shell a polyester casting from a wax original.

PLATE 12

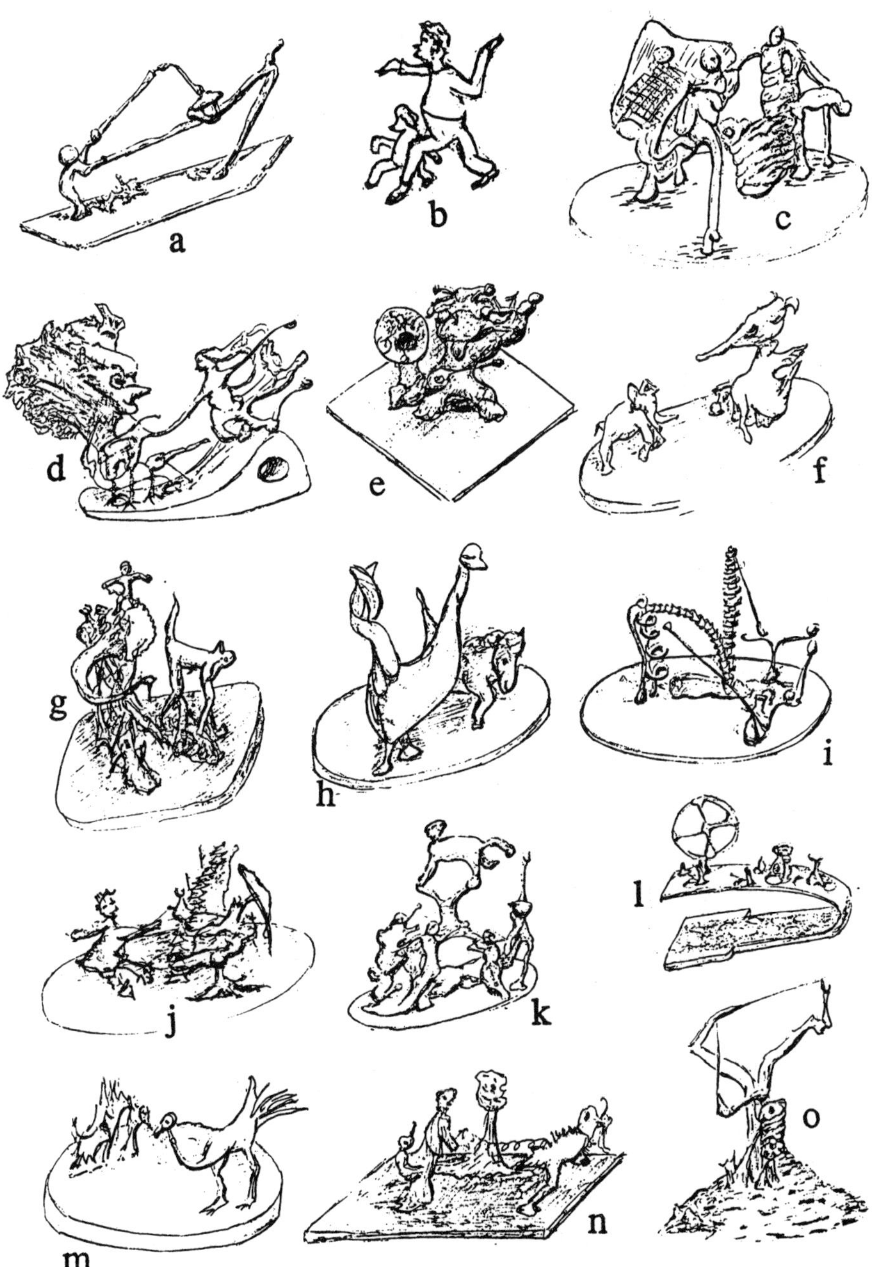

PLATE 13

MUSICAL-PUN SCULPTURES

a. *The Holey C*, a pun on The Holy See, the Vatican. 0.7"

b. *Vibraphone*, a play on the name of the musical instrument by this name, a member of the xylophone family. The telephone vibrates vigorously when its wire-mounted handle is flicked. Phone: 0.6; handle and wire: 1.5"

c. *Jello, Dalí*, a macaronic play on the song title *Hello, Dolly*. The word *Jello* is pronounced as though it were a spanish word (thus sounding like *hello*). The piece itself is based on Dalí's painting *La Persistence de la Mémoire*. 1"

d. *Gladly, The Cross-Eyed Bear*, a play on the hymn title *Gladly, The Cross I'd Bear*. Gladly's umbrella was produced from low-density polyethylene sheet stock by means of the yoke/plug procedure (p. 83). 0.7" This piece is based on a much larger one (2.5") made in ceramic clay, bisque fired, and painted with acrylic paints.

e. *The Clef in the Rock*, a play on the hymn title *The Cleft In The Rock* or the second line in *The Rock of Ages* (...cleft for me). In this piece, a treble clef is embedded in a fissure cut in the "rock" with the electric-loop tool, the rock built up from pieces of sheet stock fused together to form a near-solid mass. Color blended in from the printing on the scrap stock to simulate that of granite. 0.8"

f. *Bach's Lunch*, a play on *Box Lunch*. The initial (J.S.B.) are the telltale clue to the pun's identity. Cotton thread is used to bind the lunchbox, this being the only departure from polyethylene in its construction. 0.6"

g. *The Bell Of The Ball*, a play on the title of the waltz *The Belle Of The Ball*, written by Leroy Anderson, hence the initials *L.A.* on the ball. The ball is a glass marble, but with a little patience the purist could make it from polyethylene by means of the miniature lathe shown in Fig. 24. The bell was formed by the yoke/plug molding process (p. 83). 1.2"

h. *The Man You Ate In Sea*, a play on *The Minuet In C*. The missionary came for dinner on the South Pacific atoll shown here, the location of the shell-necklace factory seen here just before its lunch break. 0.7", (not including the palm tree). This piece is based on a much larger one (4"-tall palm tree, all else fashioned in ceramic clay but not fired). The shells used for the real necklaces, popular in the South Pacific, are those of the foraminifer *Marginopora vertebralis*, one of the larger members of this well-known

(cont. p. 190)

PLATE 13

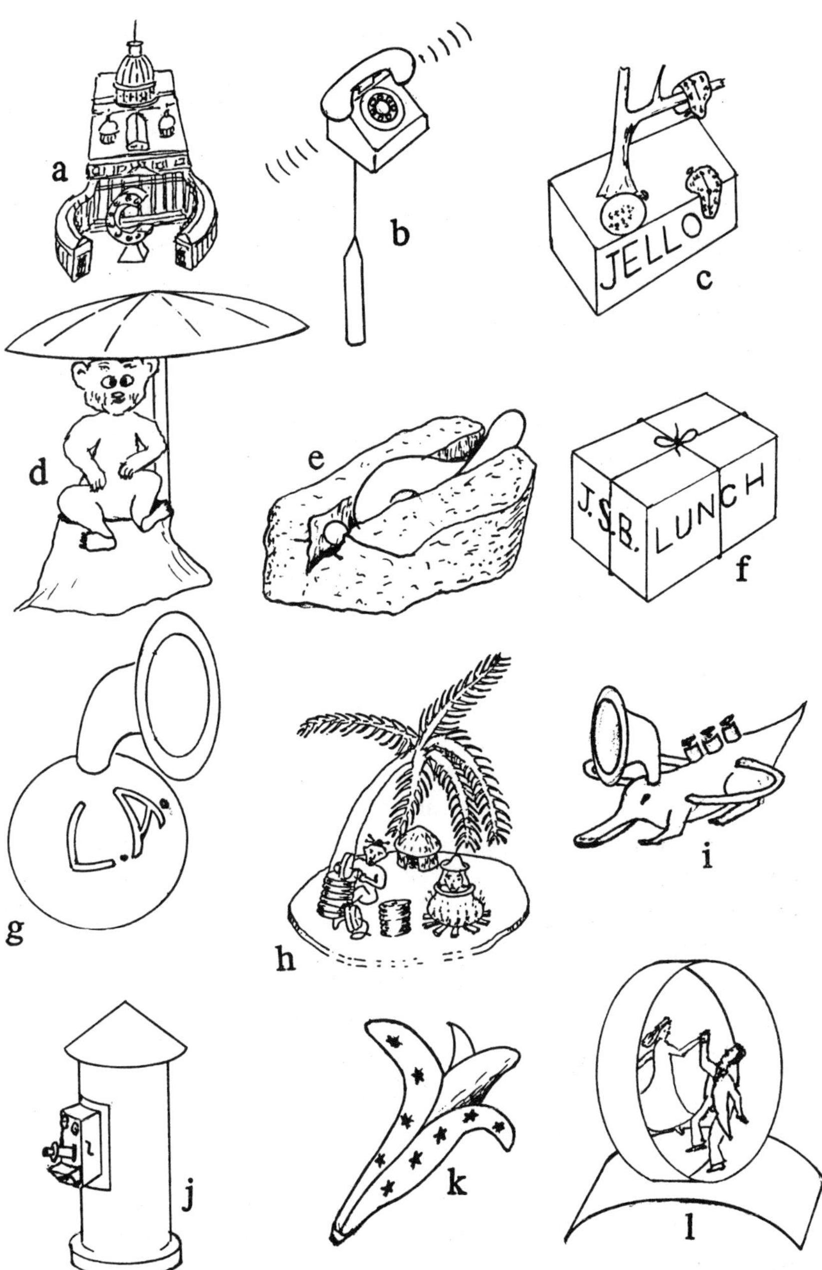

190

PLATE 14

(The pieces illustrated range from 0.2" to 1.2" in height or length.)

a. Miscellaneous turnings (i-vi) produced with the miniature lathe shown in Fig. 24, a.

b-m. Various forms that can be produced either directly on the lathe alone (b-d, g-h, and l) or basically with the lathe but then modified (after removal from the lathe) with the electric-loop tool and needles to achieve the final asymmetry characterizing each (e, f, i-k, and m).

n-v. Pieces produced either on the lathe in its conventional orientation (n, r, and u) or in its vertical (potter's wheel) position (o, p, q, s, t, and v).

Various of the above forms (particularly e, f, h-k, and l) would lend themselves to the construction of jewelry or ornaments to be worn.

PLATE 13
(Explanation continued from p. 188)

group of usually-microscopic protozoans.

i. *Hornithorhynchus*, a play on the scientific (generic) name of the duck-billed platypus, *Ornithorhynchus*, here rigged out as a musical instrument of the brass-wind family, complete with the bell, tubes, and valves normally found in such instruments. 1"

j. *Silophone*, a play on the name of the musical instrument, *xylophone*. The silo here made from sheet polyethylene, heated and rolled around a cold (wooden) mandrel. Roof of silo molded by the yoke/plug procedure. Telephone modelled entirely with polyethylene. Silo: 1.8", telephone: 0.7".

k. *The Star-Spangled Banana*, a play on our national anthem's title, The Star-Spangled Banner. Stars are paper cutouts. 1.1"

l. *Dance Band*, a play on the musical ensemble used for dances. Figures on twisted-wire armatures, set in a segment cut from a knurled-plastic cap from a juice bottle. 0.7"

PLATE 14

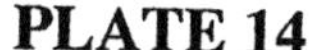

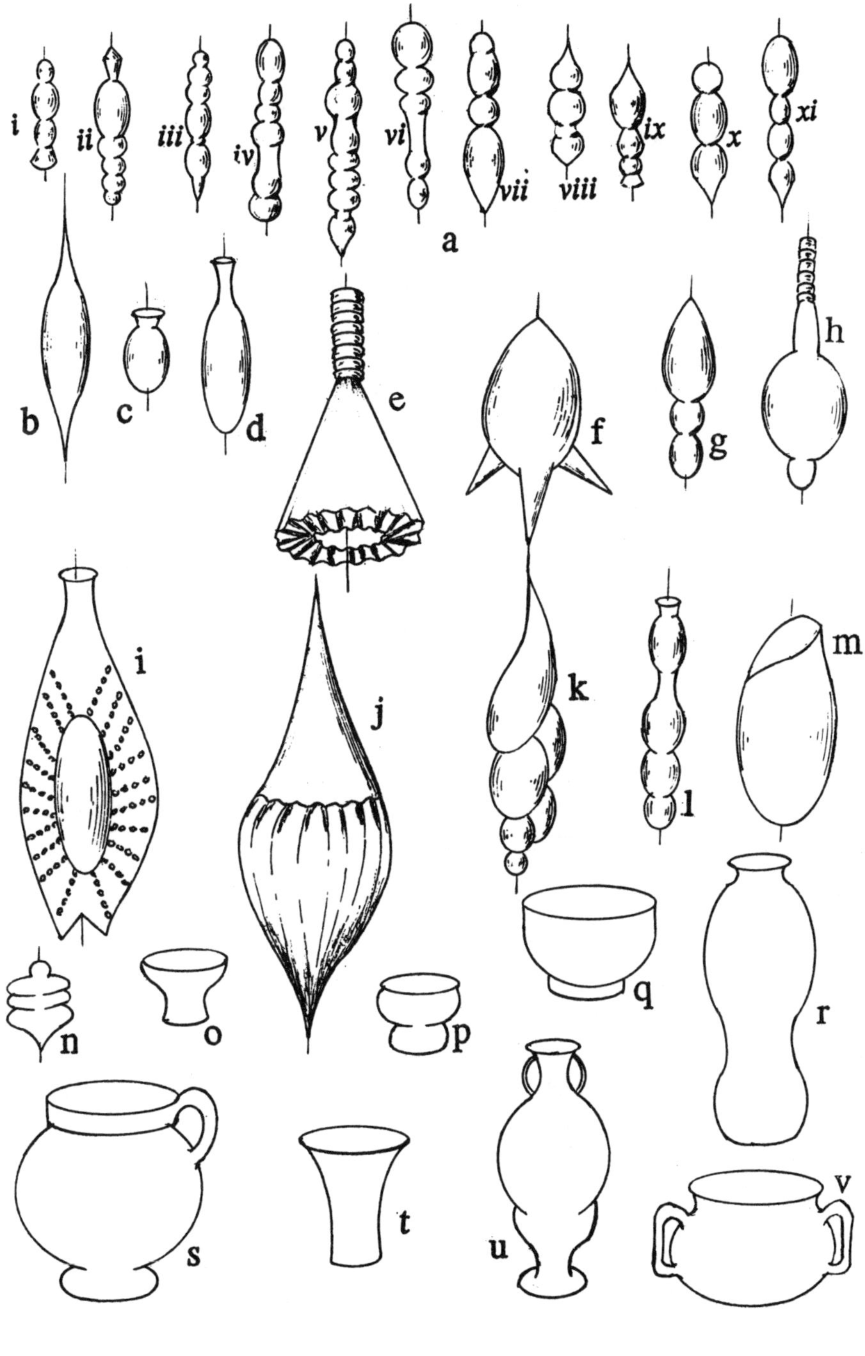

GLOSSARY AND EXPLANATION OF
USAGE OF TERMS IN THE BOOK

acatalectic- a term used in prosody to indicate that the terminal metric unit (foot) is not defective; contrasted with *catalectic.*

additive sculpture- modelling.

affective- influencing feelings or evoking emotional response.

ailurophilia (aelurophilia)- a love or affection for cats.

aleatory- depending upon chance, luck, uncertain or unpredictably erratic events or conditions.

anabolism- constructive metabolism.

anlage (pl. anlagen; from German *Anlage,-n)-* a biological term (embryological) applied to the earliest recognizable gathering of cells from which a given structure or organ will develop; here applied by analogy to the earliest recognizable shape (s) differentiated by the craftsman or artist from a blob (*urschleim,* q.v.) of molten thermoplastic material.

Angst- (German) anxiety.

autarchic- self sufficient.

banausic- utilitarian; practical(like an artisan), as opposed to intellectual.

bozzetto- a small, three-dimensional sketch made in preparation for a more finished (and usually larger) sculptural work.

calender- to press flat and smooth, usually between rollers.

cameo- a gem, sculpture or carving in relief, contrasted with *intaglio,* q.v.; when carved from natural stone having layered colors, the relief figure often stands against a background of contrasting color.

catabolism- destructive metabolism.

catachresis- misuse of a word, either intentional or not.

celadon- a color (pale-green), potter's glaze, or a piece of pottery having such color or glaze.

clepsydral- referring to a water-clock, but by extension sometimes used (incorrectly from an etymological viewpoint) to describe the shape of a sand-clock (hourglass).

concretion- here used in the geological sense of a hard, compact mass, spherical or irregular in shape, resulting from the tendency of material to gather around some central particle or matter.

corbelled (corbellated)- an architectural term, applied to a supporting projection from a wall face.

cyanoacrylate glue- powerful adhesive (super glue) made from a liquid acrylate monomer that is readily polymerized by surface-active anions.

cybernetics- science of communication and control theory (esp. automatic).

dialectic- intellectual investigation through dialogue (discussion and reasoning).

divisional sculpture- sculpture produced by division or subdivision of a larger mass into numerous smaller ones.

dop stick- a short stick or rod to which a gemstone is glued for grinding or polishing.

eclecticism- characterized by selecting, using, or relying upon materials or ideas from a variety of sources.

emboss- to raise (i.e., in relief) a design or figure from a surface.

empathic- intuitively capable of vicariously experiencing another's feelings.

engrave- to produce a design or figure, usually by incising, or cutting into, the material, leaving the background higher, as contrasted with embossed, although by extension it may be synonymized with *emboss*.

experimental aesthetics- the experimental study of the psychological explanation of aesthetic feelings.

espieglerie- frolicsomeness.

extradotal- separate property, not included in the wife's dowry.

faecal pellets- excreta (principally of marine invertebrates) found in marine sediments or as fossils in sedimentary rocks.

fimbriate- fringed.

foraminifera- one-celled animals (many microscopic, but some visible to the naked eye and of impressive dimensions for unicellular creatures), principally marine, that produce lovely shells in great variety; important also as fossils in reconstructing past environmental conditions and in correlating geological strata, particularly in petroleum exploration.

frottage- (French) rubbing; the process of transferring the impressions of relief features from a surface to paper by pressing the two tightly together and rubbing the back of the paper with a pencil.

glauconite- an earthy dark-green mineral (iron potassium silicate) occurring in greensands; often found as a replacement in faecal pellets *(q.v.)*.

glyptic- the art of carving generally; engraving (gemstones particularly).

hexameter- (poetry) a verse having six metrical feet.

HDPE- abbreviation for high-density polyethylene, one of the most common of the polyolephin thermoplastics.

intaglio- a figure or design cut into or hollowed out in a gemstone or other substance. Contrasted with *cameo*, q.v.

iwan- (Arabic architecture) a shallow hall or large porch with barrel vault.

LDPE- abbreviation for low-density polyethylene.

mandrel- a bar, shaft, spindle or axle to which a work piece is attached for machining or otherwise working.

maquette- a small preliminary model of a projected piece of art or architecture.

medallion- a large medal, as contrasted with medal (usually small) or medalet (a small medal).

metonomy- the use of one word in the place of another suggested by it; by extension, also applied to the transfer of an idea or expression (physical, emotional, etc.) from one art form (e.g., music) to another (e.g., sculpture).

mimesis- mimicry, imitation, used particularly here in reference to that achieved between different forms of artistic expression, e.g., music and dance, sculpture and music, music and poetry.

mineral segregation- the concentration or nonuniform arrangement of minerals in a molten or solidified rock mass or in sediment as a consequence of the chemical rearrangement of minor constituents.

model- to build up or shape a plastic material (clay,wax, etc.); contrasted with *sculpt,* q.v.

multiplicative sculpture- replication or multiplication of the elements or components of a sculptural piece, as through the use of molds and casts.

numen- the spiritual essence or presiding spirit of something.

paranoic-critical method- Salvatore Dalí's "spontaneous method of irrational cognition based upon the interpretative-critical associations of the phenomena of delirium".

paraphernal- the separate personal or real property of a married woman; (see also *extradotal*).

pegmatite- a coarse variety of granite, often injected (molten) as a tabular mass (dyke) into a fissure.

pendentives- a concave triangular support for a dome over a square area.

PETE- abbreviation for polyethylene terephthallate.

PP- abbreviation for polypropylene.

PS- abbreviation for polystyrene.

polymerization- the change of one substance into another (by the union of two or more molecules of the same kind) having the same elements and elemental proportions but higher in molecular weight and exhibiting different properties.

postscenium- a frame set behind the screen of a shadow theater to define the area of screen to be illuminated.

postiche- added unnecessarily, or superfluously (and particularly inappropriately) to an otherwised finished work; an unnecessary or inessential appendix.

protean- variable or changeable in shape, by extension (particularly in biology) simple or primitive.

punstrument- a novel musical instrument or device based on a musical pun and usually played by means of a miniature harmonica concealed in or attached to the punstrument.

pyrogenics- (geology) arising or produced through the action of heat and pressure, as igneous rocks.

pyrognostics- (geology and mineralogy) mineral analysis through a study of the reaction of the substance to the heat and flame of a blowpipe.

pyrophile- one who loves fire.

pyrophobe- one who hates fire.

sculpt- to carve or remove; applied to the act of sculpture in its strict sense, as opposed to modelling.

silicoflagellates- any one of a group of flagellated protozoans capable of secreting siliceous skeletons.

solipsism- absolute egoism and, by extension, egotism.

solipsist- one characterized by solipsism.

squinch- (architecture) a supporting structure, such as an arch, across an interior or re-entrant angle.

stalactite- (geology) a mineral deposit (usually calcium carbonate) hanging from a cave roof, in form like an icicle and produced by the dripping of mineral-rich groundwater. In architectural sculpture, the term is applied to such ceiling ornamentation as that produced by clusters of corebelled squinches, as seen, e.g., in the Court of Lions in the Alhambra (Granada, Spain).

stalagmite- a mineral deposit (usually calcium carbonate) produced on the floor of a cave, in form like an inverted stalactite, and produced, like it, by the dripping of mineral-rich groundwater.

subtractive sculpture- involving the removal of material from a mass; carving; sculpture in the strict sense (as opposed to *modelling*).

Symbolist movement- late 19th-century French movement to unify the arts and blend the sensory functions.

synaesthesia- the simultaneous perception of dissimilar stimuli.

synaesthetics- the comparison or fusion of aesthetic reactions from different artistic regimens.

thermoplastic- the property of having a molecular structure that will repeatedly soften when heated and harden when cooled, the original characteristics of the material retained; also, any substance having this property. Note: the fact of the word being used as both an adjective and a noun makes it ambiguous and can lead to misunderstanding, but the plural form (thermoplastics) is a noun only and is, thus, unambiguous, even when used adjectivally, as in the title of the present book.

thermosetting- having the property of solidifying when heated under pressure but not capable of being remelted or remolded without losing its original characteristics.

topoi- (sing.:*topos*) stock themes, topics, or points of reference in rhetoric and, by extension, in musical analysis.

toroidally-wound transformer- one produced by winding a number of current-bearing coils of wire around a doughnut-shaped magnet, the magnetic field essentially contained within it by nature of the magnetic material.

urschleim- (German *Urschleim*) the primitive slime or ooze from which life was earlier speculated by biologists to have arisen. Applied in the present work to the undifferentiated mass or blob of plastic from which sculptural forms (in their early developmental stage here termed *anlagen*, q.v.) are subsequently to be developed.

ziggurat- (architecture) an ancient (Mesopotamian) pyramidal temple tower.

SCULPTORS, ARTISTS,
AND WORKS REFERRED TO

(The page reference for each is given in parentheses. Specific works italicized, general ones not.)

<u>Sculptor/Artist</u>	<u>Name or type of work</u>
Allport, Alan	book on paper sculpture (6)
Arp, Jean	*Human Concretions* (128,157-8), *Landmark* (158), *Ptolemy I* (118, 158), *Star* (118, 158)
Boccioni, Umberto	*Anti-Graceful (Mother)* (131, 159), *Unique Forms of Continuity in Space* (131, 159)
Brancusi, Constantin	*Bird in Space* (130, 131, 155, 158), *Bird* (158), *Mademoiselle Pogany* (vi, 158)
Canova, Antonio	*Meekness* (159)
Cascella, Andrea	*The White Bird* (159)
Cozens, Alexander	book on blot paintings (119)
Dalí, Salvatore	*paranoic-critical* method (119, 124) *Architectonic Angelus of Millet* (159)
Ernst, Max	frottage (5,118)
Escher, Mauritz	(118)
Giotto	dejected monk, detail from his *The Vision of the Burning Chariot* (vi)
Haeckel, Ernst	book of biological drawings (118, 158)
Hajdu, Etienne	*Delphine* (116, 159)
Hepworth, Barbara	monumental works (128, 155), pierced forms (157, 159, 160)
Jancic, Olga	*Double Form* (157)
Kandinsky, Wasilly	*Sky Blue* (108)
Kowalski, Piotr	*Calotte 4* (157)
Leonardo da Vinci	chance images as artistic sources (118)
Limbourg brothers (Pol, Hennequin, and Herman de)	*Les Très Riches Heures du Duc de Berry* (vi)
Lipchitz, Jacques	semi-automatic sculpture (126), *Benediction I* (157)
Martins, Maria	*Rituel du Rhythm* (41, 118, 159)
Matisse, Henri	*Tiaré with Necklace* (157, 159)
Michelangelo Buonarroti	carving vs. modelling in sculpture (30)

OUR SCULPTURAL HERITAGE

We all owe a lot to the sculptors of old.
In pursuit of their art they were mighty and bold;
They knew how to chisel and whack away stone,
And had not a fear of Nature full blown.

Their religion was Truth, their god was Fact,
And both they portrayed with abundance of tact.
You knew where you were when you looked at their work;
As craftsmen, their standards they never would shirk.

True craftsmen they were: they mastered their tools,
But always adhered strictly to the rules.
Flim-flam and four-flushing they all did abhore;
To capture true Beauty, they'd ask for no more.

Indebted to them, should we not try
Our sights to set on goals that are high,
And leave our descendants works just as bold
As those of the fearless sculptors of old?

A MUSICAL APPENDIX

As a mild exercise in experimental aesthetics and as a study in synaesthesia, the poems scattered through the text are here supplemented by a musical appendix. This appendix, an extension of the basic search for the numen of miniature thermoplastics sculpture begun in Part Four (p.111), leads to a consideration of the relation between music and miniature thermoplastics sculpture. Since the metonomy and catachresis so often characteristic of synaesthesial dialectic can be an additional complication for those of us unaccustomed to philosophical and psychological attempts at analyzing and describing, as a comparative study, the concomitant moods and sensations elicited by juxtaposed works from two different fields of artistic endeavor, I shall omit detailed discussion and leave an evaluation of the fitness or unfitness of the juxtaposed pairs of sculptural figures and musical scores to the viewer, reader, and listener.

The Symbolist movement in late 19th century France, under the guidance of Verlaine, Mallarmé, and the Belgian Maeterlinck, gave earnest consideration to the relationship between poetry and art, while Stumpf (1890) probed deeply the psychology of musical tones, attempting to depict and describe the nature of the moods excited or stimulated by music. More recently, Allanbrook (1983) has anatomized the relation between the gestures of musical rhythm and dramatic art. In her detailed analysis of the *Allegro* movement of Mozart's F major piano sonata (K. 332), she carefully identified the principal musical motifs that, because they had implicit connection with ordinary human posture (movement, gesture) could be directly related to these in the listener's own apperceptive endeavors, that they are, in effect "subliminally referential" as one experiences the aural sensations. If one accepts the thesis, early proposed by Aristotle and often reinforced through successive centuries down to the present day, that external appearances and actions are clues to internal features of the soul's character, then it is easy to accept the suggestion that rhythmic gestures in music (i.e., motifs) choreograph the body movements of the actors and, thus, tell much of the character being portrayed in the opera.

The history of such comparative studies of the aesthetics of two different arts is a long one indeed, its literature rich and varied. It would not be surprising to find that in our modern world students philosophically interested in or actively and creatively involved with the cybernetics explosion and its shock waves (to the publishing and advertising industries, for example), the evolution of electronic communication, and the exploitation of the vast potential of multimedial presentation might find worthwhile an investigation of the long history of and vast literature on comparative aesthetics in general and synaesthesial phenomena in particular.

Certainly, in the light of the intellectual climate in artistic and belle-lettristic circles near the close of the 19th century, and with the writings and thinking of the time as undoubtedly a part of his own apperceptive mass, it is not surprising that Kandinsky came soon thereafter (1914) to feel that in art "relationships are not necessarily ones of outward form, but are founded on an inner sympathy of meaning". He was, in fact, merely echoing thoughts of a long lineage of philosophical predecessors. In his painting, however, Kandinsky consciously strove to depict—often by non-objective means—an inner harmony that would be as powerful in evoking a deep emotional or spiritual response from a viewer as the combined melodic, harmonic, and rhythmic delights of a musical composition can be in revealing a deep-seated spiritual harmony within a listener. Sadler (1977), in the introduction to his translation of Kandinksy's (1914) book, observed that "The power of music to give expression without representation is its noblest possession". The goal of Kandinsky and of his non-objective colleagues would seem to have been to develope similar power in their painting.

In attempting to discover relationships between music and the sculptural pieces illustrated in the present book, it was immediately obvious that greater reliance must be placed on mere form or line than on color, little color having been used in the actual pieces (in the attempt to emulate ivory and white marble) and none, of course, in the corresponding line drawings. Even so, a word of caution should be inserted about the ease with which an observer can mimic in his own reactions patterns observed in an object and then imaginatively transfer these and interpret them as part of the object's own dynamic properties, a striking example of empathy, just one of the many pitfalls to interpretation awaiting the unwary or overeager experimentalist.

The task of comparing the signs or symbols of a musical motif—many of which are essentially "empty", or devoid of emotional inference—with the line and form of an objective sculptural representation of part of nature is, of course, more difficult than would be that of simply comparing two works of objective art, but since most of the illustrated pieces are, in fact, naturalistic (objective) rather than symbolic (non-objective), our search for spiritual harmony is necessarily more superficial and less penetrating than would be that of a non-objective artist desiring to depict and evoke harmony without any resort at all to representation. The primitively or crudely naturalistic character of the illustrated pieces merely reflects the technical and artistic incompetence of a novice; a few of them, even so, do tend to wander from the track of naturalism and even verge on the surrealistic or worse. Most of them, in addition to being essentially objective, correspond, furthermore, to what Kandinsky would call *melodic*, i.e., simple, rather than *symphonic* (complex) compositions had they been paintings rather than sculptures, although, here again, a few tend at least toward polyphonic or contrapuntal expression while falling far short of the grandeur of a symphonic complex. Like my vocabulary of musical *topoi*, my sculptural vocabulary is probably grotesquely inadequate in its untutored puerility, but perhaps enough recognizable elements exist in each to permit at least rudimentary comparison.

In order to suggest certain parallels between the mood of the sculpture and that of the music (hopefully demonstrated by the successful evocation of an emotional concurrence from the viewer when the music is played on the piano), I have affixed sketches of what I consider appropriately suggestive sculptural pieces to the various musical strains or excerpts where the line, movement, weight, balance, and the overall harmony or discord of the sculptural piece seem to me to be in spiritual accord with those of the music. The juxtaposition in this appendix of sculptural figures and musical scores is predicated, on the one hand, on the experimentally demonstrable empathic value of mere line and gesture in evoking emotional response and, on the other, on the widely recognized power of music in such evocation. To the judgement of the viewer, reader, and listener, however, is left the appraisal of the degree of agreement in response evoked by the visual and aural stimuli, the resemblance or harmony between the appeal of the two to affective experience. Far from presuming to suggest anything deeper than a mere superficial spiritual or emotional sympathy or empathy between two sets of quantitatively and objectively ineffable aesthetic values and impressions, I would merely propose that the music be used as a background for viewing or displaying the actual sculptural pieces or, more ambitiously, as incidental music to accompany a shadow-play presentation by figures and scenes in the miniature shadow theatre shown in Fig. 42, a (p. 145).

The sources of the musical examples (all excerpts from original piano compositions by the author) are indicated on the next page. The titles of three of the excerpts (*) have been supplied or modified specifically for the purpose of indicating more accurately the spirit of the attached sculptural illustrations.

Page	Title	Source Composition		Copyright
203	Invitation (to the dance*)	*Pieces for the Piano*	Opus 2, no. 3.	1987
204	Tranquillité*	*Lyric Suite*	Opus 2, no. 2.	1984
205	The Nonchalant Rag	*Concert Dances*	Opus 5, no. 2	1989
206	Frolic (for left hand alone)	*Pieces for the Piano*	Opus 2, no. 3	1987
207	Valse Nostalgique	same		
208	Caprice	same		
209	Improvisations Protistologique	*Album Maritime*	Opus. 5, no. 1	1989
210	Scherzo*	*Lyric Suite*	Opus 2, no. 2.	1984
211	Christmas Melody I	*Christmas Album*	Opus 4, no. 1.	1986
212	Postlude	same		

CATACHRESIS AND SYNAESTHETICS

Experimental aesthetics for some
Is a field devoid of allure,
But for those who to its charms succumb
There is, I fear, but one known cure:

Continued study of relations obscure,
Deriving from inference in part subliminal,
But yielding results that seem to endure
Attacks by eloquence abominable.

For those to synaesthetics drawn,
However experimental their bent,
Conclusions weird and wild they'll spawn
Though not, perhaps, their true intent.

Just happenstance, I suppose you might say;
Things didn't go as they might have liked.
Unhappily, their stance often gives them away;
You might even say they were psyched.

And why should psychologists not be fooled
By empathy and catachresis?
Indeed, we know they weren't all schooled
To deal with these in their doctoral thesis.

INVITATION (TO THE DANCE)

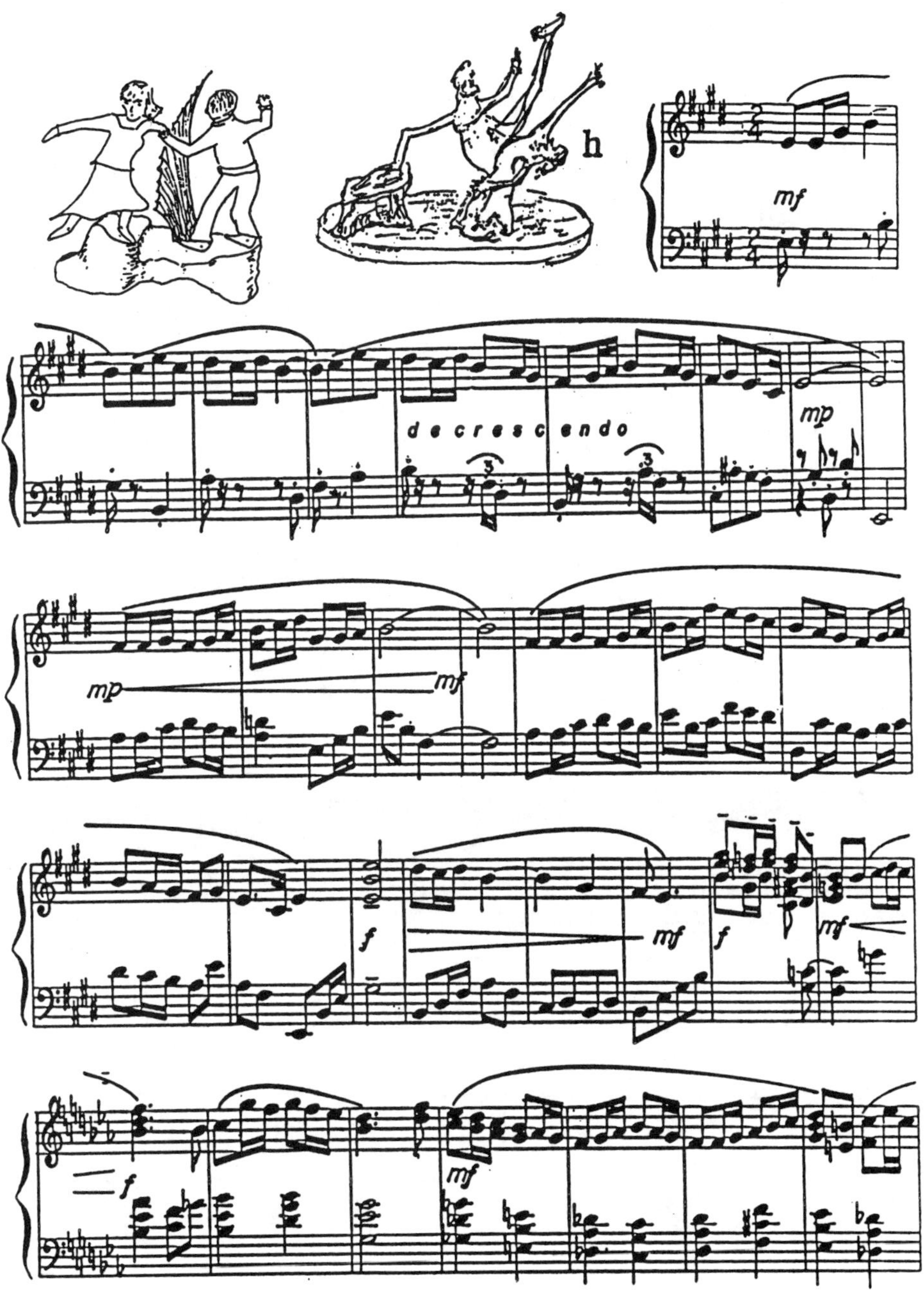

TRANQUILLITÉ

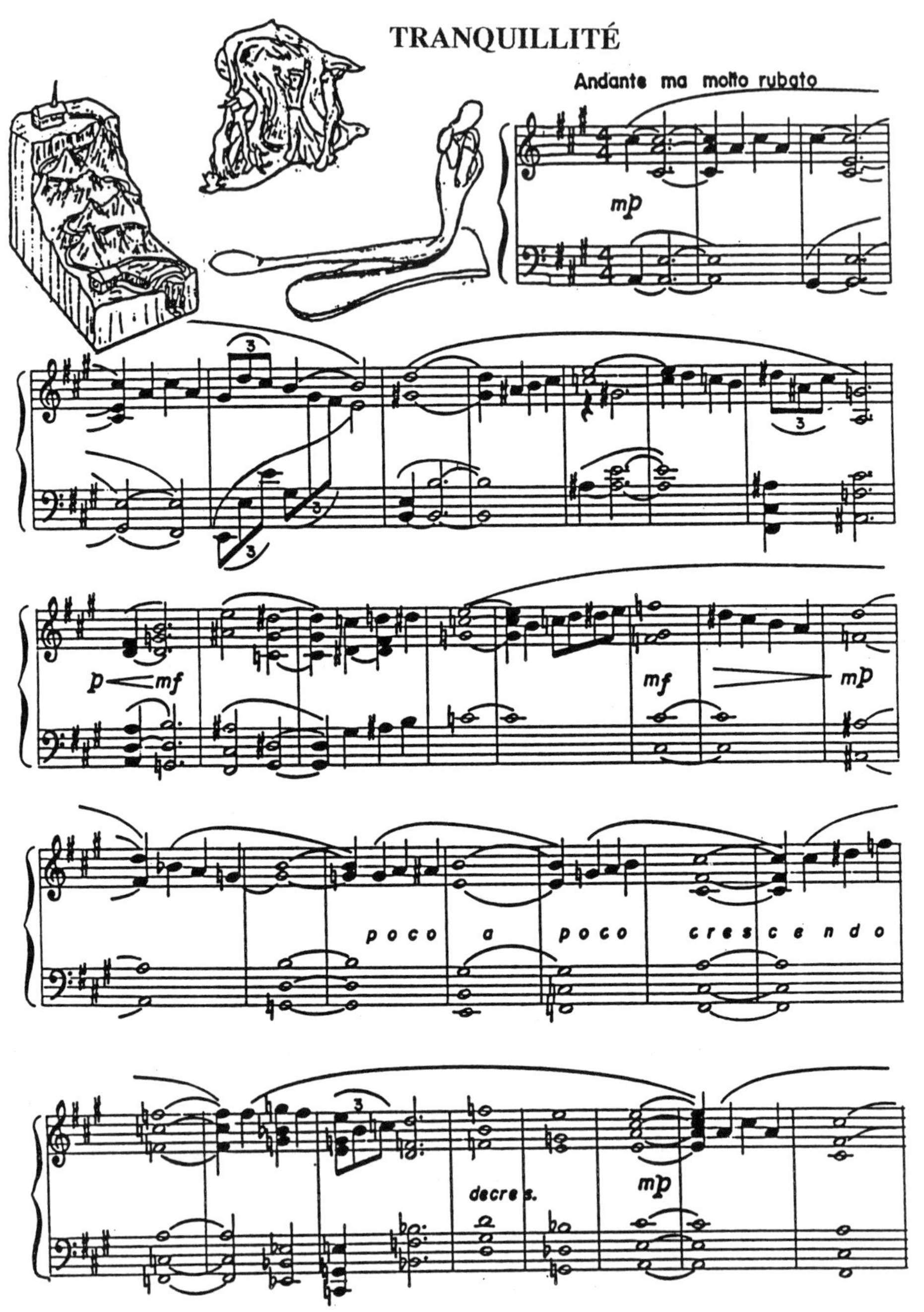

THE NONCHALANT RAG

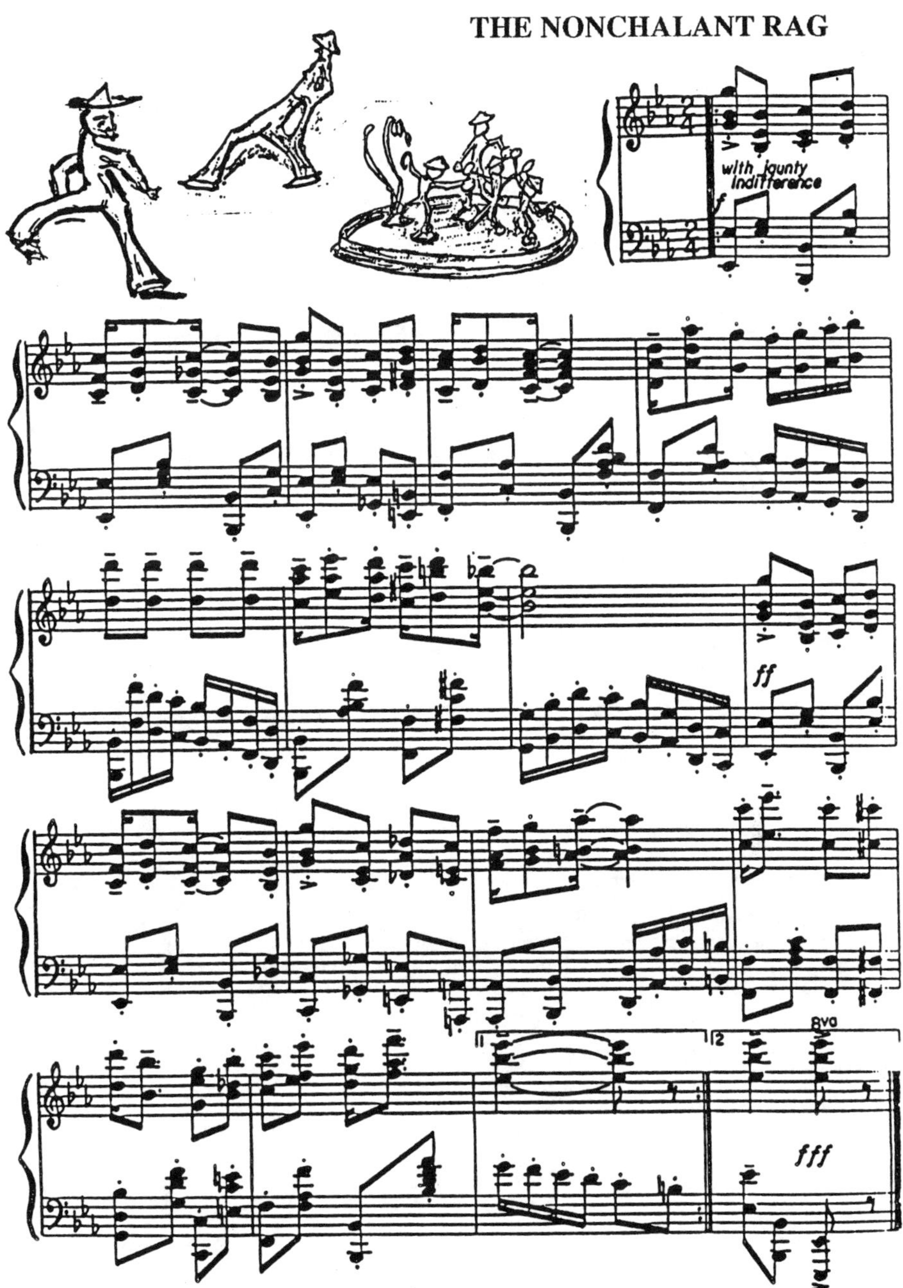

FROLIC
(Left Hand Alone)

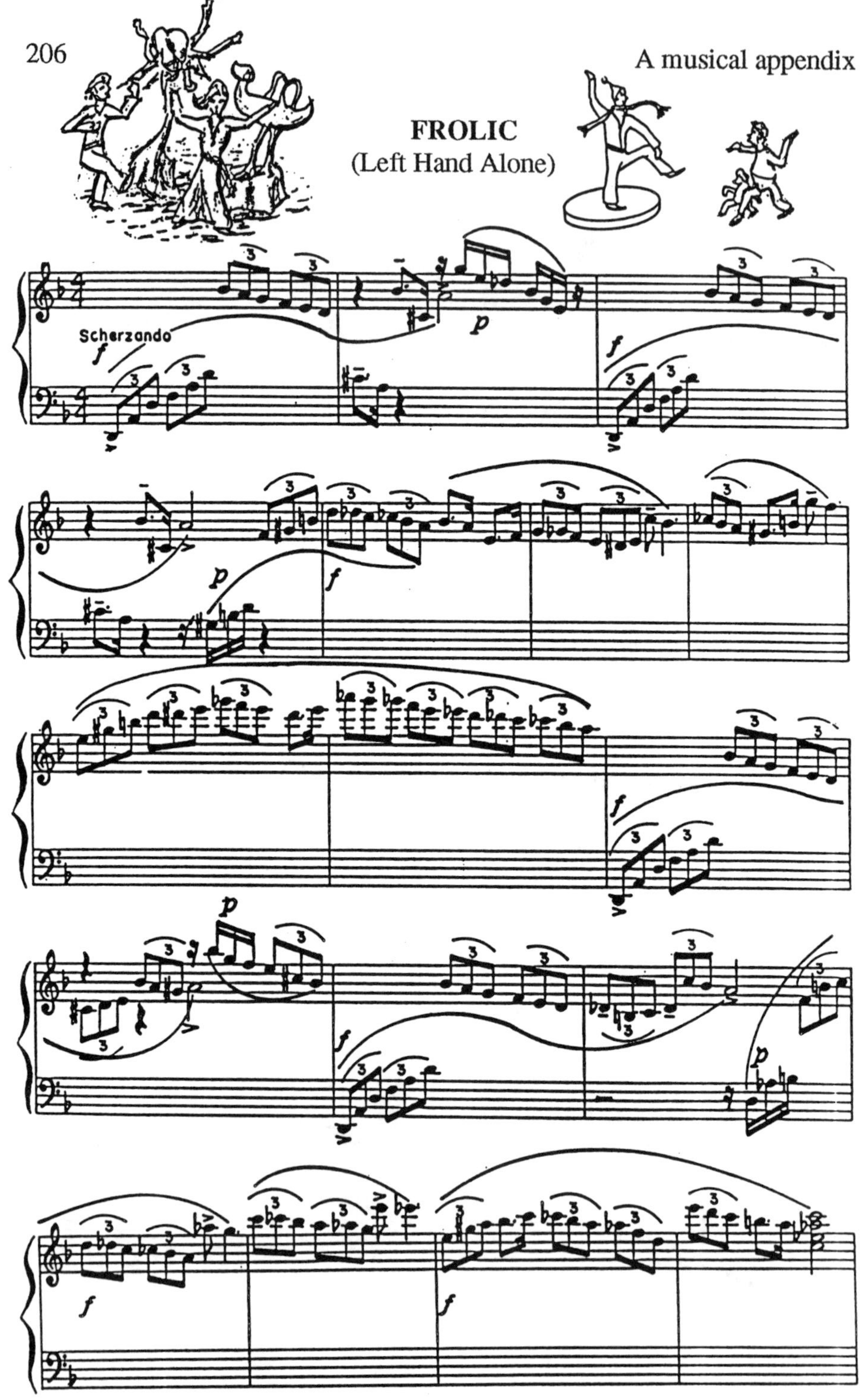

VALSE NOSTALGIQUE

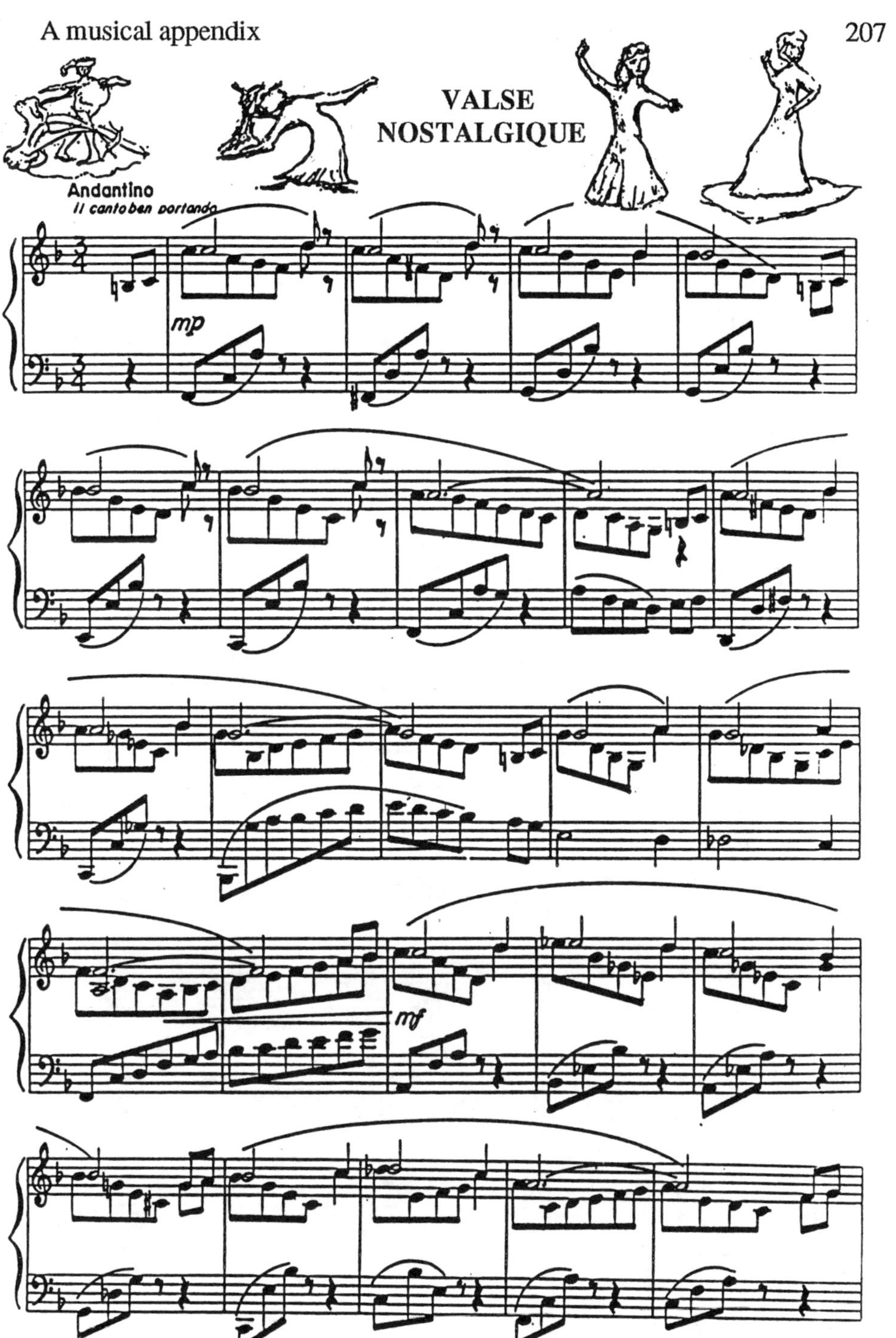

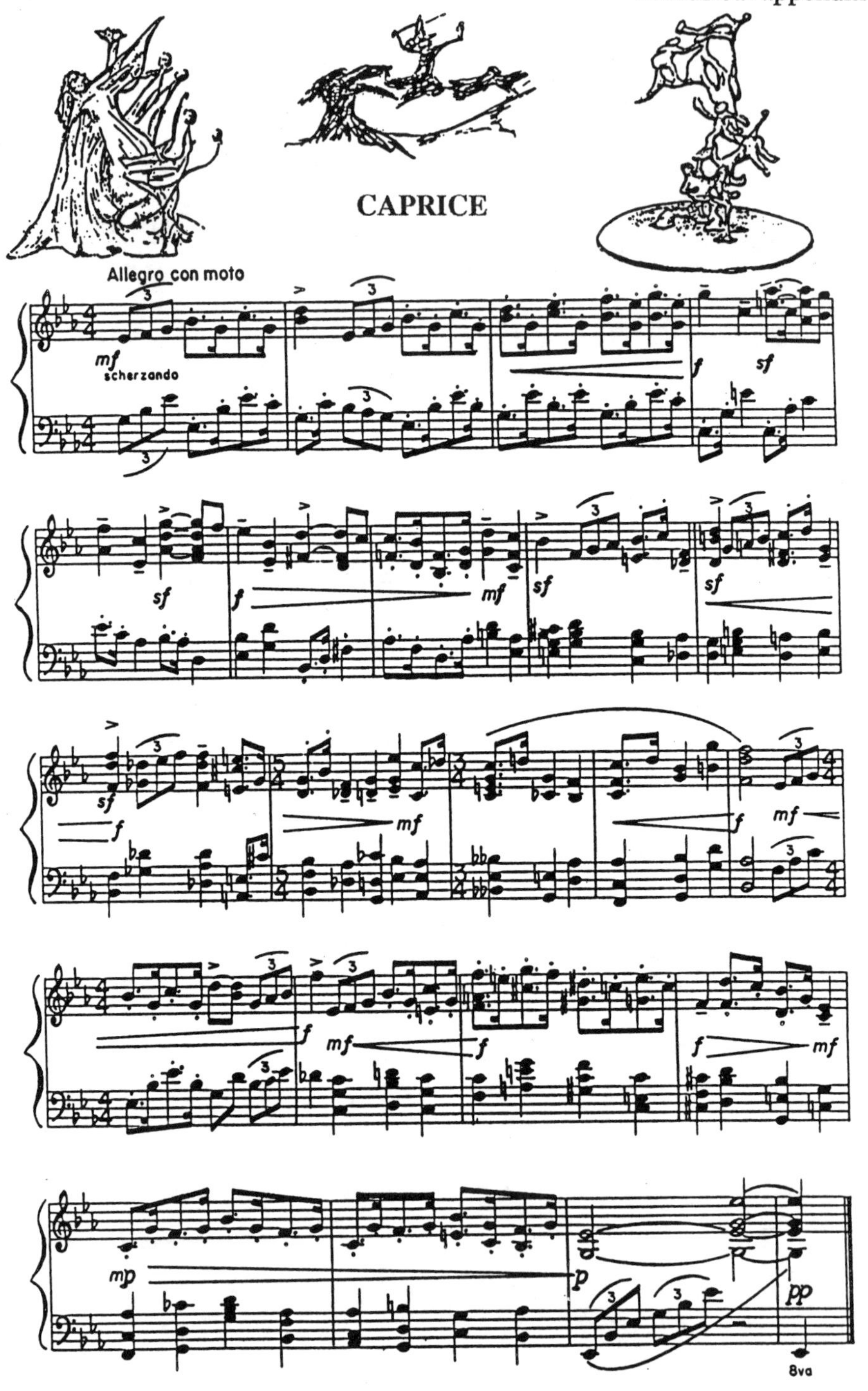

CAPRICE
Allegro con moto
mf
scherzando
sf
f
sf
f
mf
sf
sf
sf
f
mf
f
mf
f
mf
f
mf
f
mf
mp
p
pp
8va

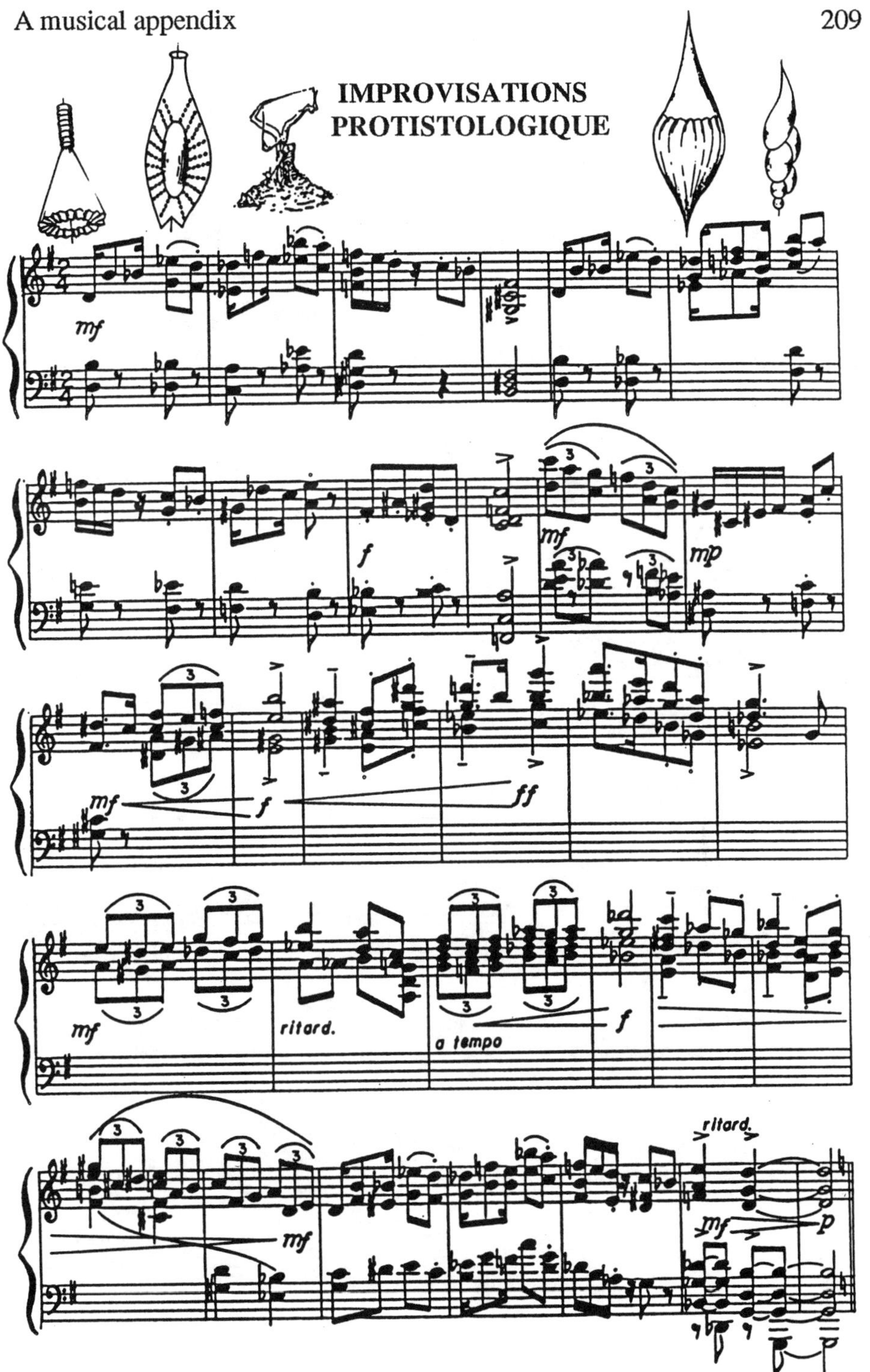
IMPROVISATIONS
PROTISTOLOGIQUE
mf
f
mf
mp
mf
f
ff
mf
ritard.
a tempo
f
ritard.
mf
mf
p

SCHERZO

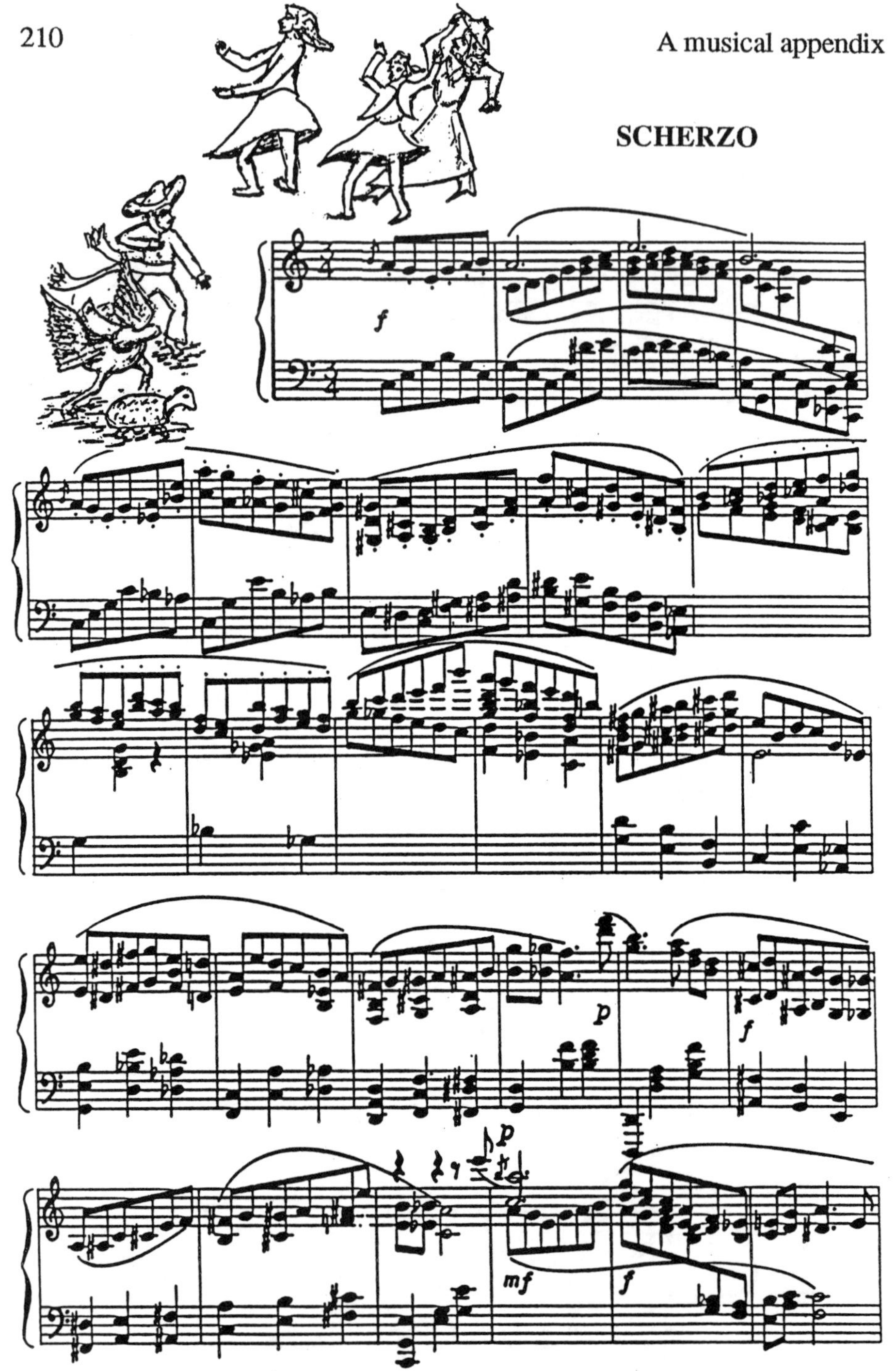

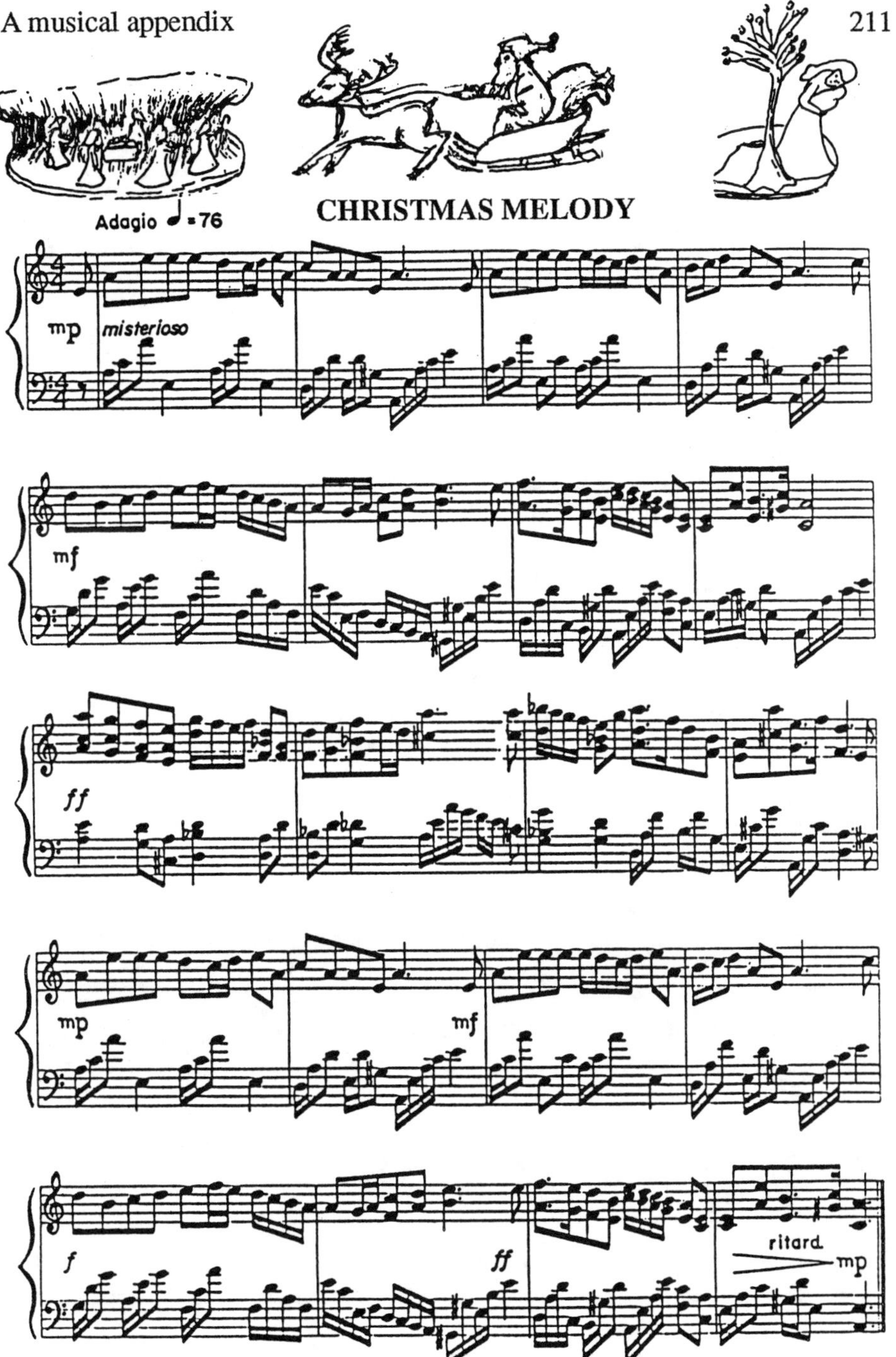
Adagio ♩=76
CHRISTMAS MELODY
mp misterioso
mf
ff
mp mf
f ff ritard. mp

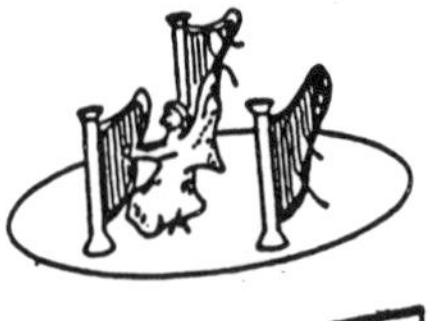

POSTLUDE

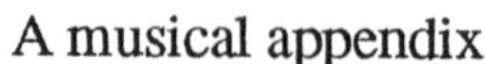

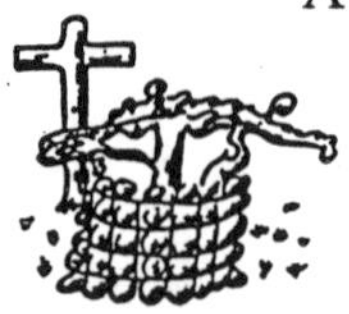

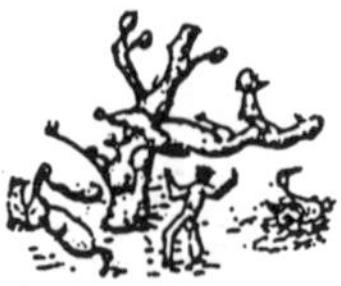

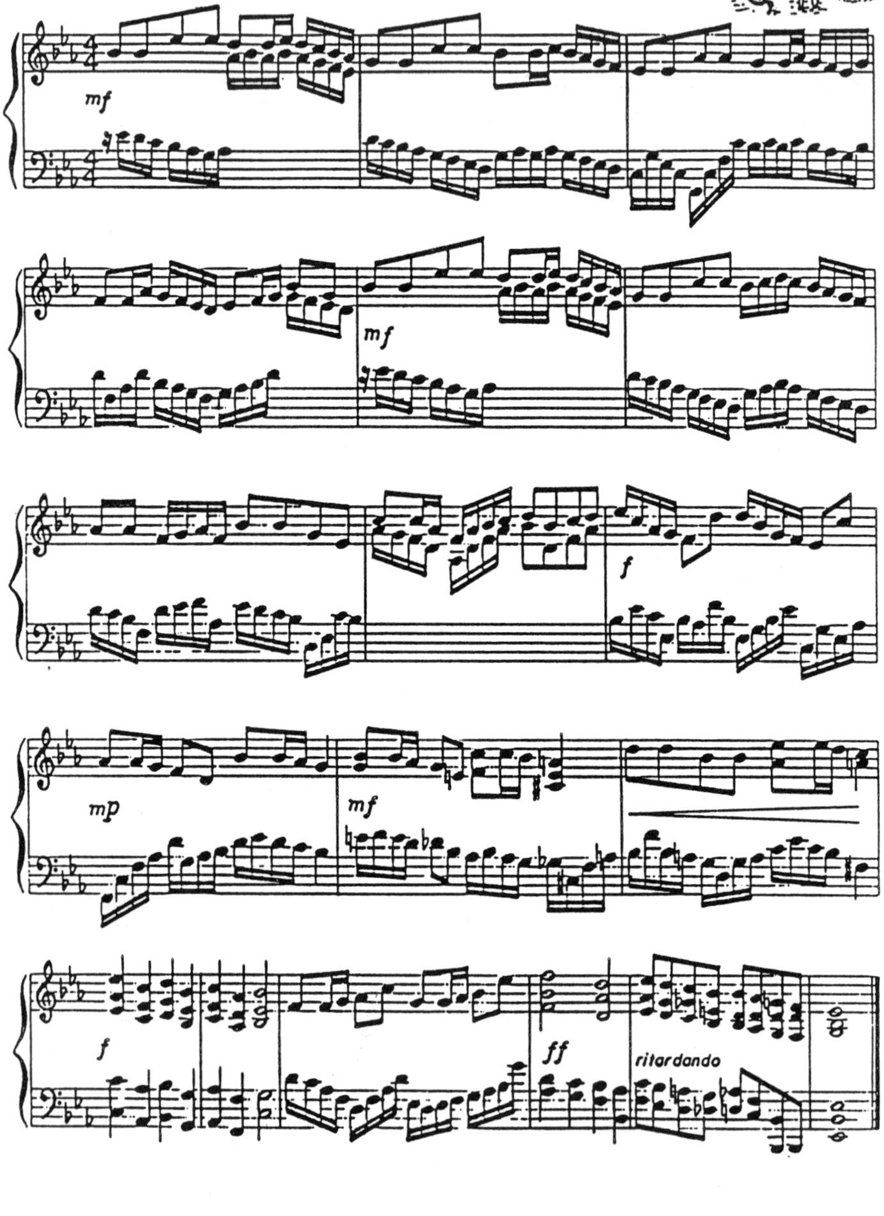

LITERATURE CITED

Ades, Dawn, 1982. *Dalí*. London, Thames and Hudson. A sensitively penetrating study of one of Surrealism's most imaginative geniuses.

Allanbrook, Wye J., 1983. *Rhythmic gesture in Mozart: Le nozze di Figaro and Don Giovanni*. Chicago, U. of Chicago Press. An extensive study of the role of musical *topoi* in operatic expression and mimesis.

Allport, Alan, 1971. *Paper sculpture*. London, Pelham Books. A delightful manual for the dexterous craftsperson who wishes to explore in depth the potential of paper as a three-dimensional, or sculptural, medium.

Arnold, Zach, 1994. *Musical punstruments*. Pacific Grove, Boxwood Press. Source of musical puns and the design and construction of novel musical devices (punstruments) based on these, including some on which miniature sculptural pieces have been based.

Bool, F.H. *et al.*, 1992. *M.C.Escher*. New York, Harry Abrams. A biography of this unusual artist, and a magnificent collection of the prints through which the artist is playfully making nonsense of some of the irrefutable certainties of perspective, spatial, surficial and gravitational relationships.

Clark, C. Dame, 1940. *Molding and casting*. Baltimore, Lucas. Consulted but not cited in the text. This is a valuable guide to the subject, although, of course, some the valuable modern materials were not available at the time of its writing.

Cook, R.M., 1972. *Greek art*. New York, Farrar, Straus and Giroux. A useful source book and refence guide for one who wishes, as a learning experience, to study and attempt to make miniature copies in thermoplastics of Greek sculpture, seals, and cameos.

Cozens, Alexander, 1786. *A new method for assisting the invention in drawing original compositions in landscapes*. Pioneer work on blot painting, although Leonardo had earlier suggested various unconventional sources of inspiration for artists.

Cushman, Joseph, 1948. *Foraminifera*. Cambridge, Harvard Univ. Press. This book on the delightful group of shell-forming marine microörganisms was for years the standard textbook for many university courses in micropaleontology.

Deschames, Robert and Gilles Néret, 1989. *Salvador Dalí*. Köln, Taschen. A penetrating study of the artist and a magnificent set of reproductions of his works. The miniaturist will find that many of the devices and effects Dalí used so masterfully can be adapted for use in thermoplastics sculptural pieces.

Dupont, Jacques, and Cesare Gnudi, 1979. *Gothic painting*. Geneva, Skira. A scholarly text with magnificent illustrations for the inspiration and guidance of anyone seeking to learn from and to emulate in thermoplastics the technical and artistic excellence of medieval artists.

Du Ry, Carel, 1970. *Art of Islam*. New York, Harry Abrams. A handsomely illustrated guide to the delights of corbelled squinches, stalactitic pendentives, and innumerable other structural and ornamental features of Islamic architecture from which the student of miniature thermoplastics sculpture should derive much inspiration.

Ernst, Max, 1948. *Beyond Painting*. New York. Wittenborn, Schultz. A master of the application of chance effects, as from his *frottage*, to the interpretation of hidden meanings and personal reactions.

Evers, Hans-Gerhard, 1963. *Zeugnisse du Angst in der modernen Kunst*. Darmstadt, Evers. Written in accompaniment to the exhibit of modern art held in Darmstadt in 1963.

Haeckel, Ernst, 1904. *Kunstformen der Natur*. Leipzig and Vienna, Verlag des bibliographischen Instituts. (Dover reprint: 1974 *Art forms in nature*. New York, Dover.) One hundred large-format plates of exquisitely detailed drawings of animals and plants from algae and protozoa to pitcher plants and antelopes. A design treasury for the sculptural miniaturist seeking inspiration from some of Nature's own sculptural marvels, miniature and otherwise.

Haftmann, Werner, 1965. *Painting in the Twentieth Century. vol. 1*. New York, Holt, Rinehart and Winston. A valuable and comprehensive source book for the student seeking confirmation of the close relation between painting and sculpture; presents a wide range of forms and subjects from which the miniaturist can derive inspiration and guidance.

Hohauser, Sanford, 1970. *Architectural and interior models*. 1970. New York, Van Nostrand Reinhold. A useful source of ideas for adaptation to the needs of the sculptural miniaturist, particularly in the development of pieces having architectural components.

Kandinsky, Wassily, 1914 *Über das Geistige in der Kunst*. English Translation by M.T.H. Sadler, 1977: *Concerning the spiritual in art*.New York, Dover.

Licht, Fred, 1967. *Sculpture 19th and 20th centuries*. Norwich, Jarrold and Sons. A valuable photographic collection of sculptural pieces, a convenient and compact reference source, a treasury for the student copyist, and a source of inspiration and ideas for the development of one's own original sculptural miniatures.

Lucie-Smith, Edward, 1984. *Dictionary of Art Terms*. London, Thames and Hudson. A compact and well-illustrated reference work to help such novices as I through the maze of artistic terminology.

Maclhose, L. S., 1960. See Vasari, G.,1550.

Marinetti, F. T.,1909. *Manifesto.* Paris, *Le Figaro,* Feb. 20, 1909. The "initial" manifesto of the Futurist movement by its instigator. See also Read, H. 1964, pp. 117-119, 121 for examples of the poet's powerful imagery.

Platt, Colin, 1985. *The Atlas of Medieval Man* New York, Crescent. Provides valuable perspective (particularly for such latecomers to it as I) to the history of art and architecture.

Pope, Arthur U., 1965. *Persian Architecture.* New York, G. Braziller. A scholarly, well illustrated study of the elaborate use of form and color in Persian architecture.

Read, Herbert, 1964. *A concise history of sculpture.* New York, Oxford Univ. Press. A helpful guide that, in conjunction with his *Concise history of modern painting (1974),* would serve well the student of the arts in search of a broad, sound perspective.

Redfern, W. D., 1984. *Puns.* Oxford, Blackwell. A most scholarly study of all aspects of puns and punning, a treasury of puns through the ages, a delight to read and study.

__________, 1996. *Puns: Second thoughts. Humor* 9 (2): 187-198. Berlin/ New York, de Gruyter. A brief extension of the same delightful thought-provoking scholarship and humor found in the author's earlier (1984) book.

Reynolds, Joshua, 1961 edition of 1778 original. *Discourses on art.* New York, Collier. His Discourse No. 10, on sculpture, should be of value to the would-be thermoplastics miniaturist as providing aesthetic perspective and background.

Sadler, M.T.H., 1977. see Kandinsky, Wassily.

Strong, John, et alii., 1938. *Procedures in experimental physics.* New York, Prentice-Hall. A down-to-earth guide to the mysteries and marvels of several basic technique of potential value to the experimental miniaturist anxious to broaden the horizons of thermoplastics sculptural creativity.

Stich, Sidra, 1990. *Anxious Visions: Surrealist Art.* New York, Abbeville. A comprehensive, richly illustrated study of the art of the surrealist.

Stumpf, Karl, 1890. *Tonpsychologie.* Leipzig, S Hirtzel. Vols. 1-2 (1883-1890) A scholarly philosophical study of the psychology of musical expression.

Vasari, Georgio, 1550, 1568. *Introduction* (to techniques of architecture, sculpture and painting) prefixed to his *Lives of the Artists.* Translated by L. S. Maclehose, 1960. New York, Dover. A study of the materials and basic processes or techniques in the three arts for the general reader as well as for workers in the three fields of artistic endeavor.

INDEX

Pages with illustrations are in **bold**
type. *Italicized* entries are for
foreign words and the names of
sculptural pieces, exemplified words,
musical compositions, and musical
punstruments.

A

INDEX OF FIRST LINES OF POEMS

ADIEU RÉFLÉCHI

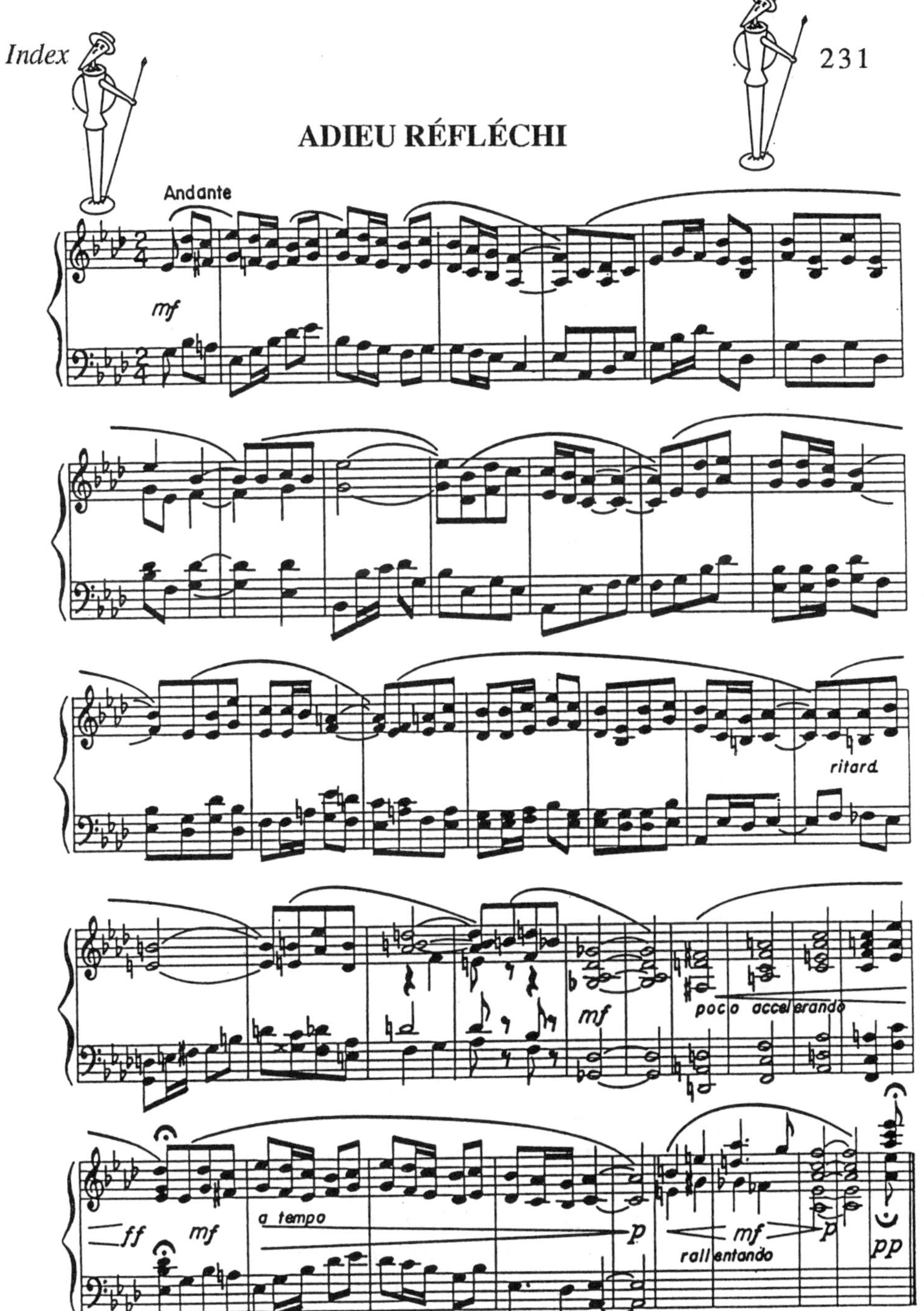

HAPPY, INDEED, IS HE OR SHE

Happy, indeed, is he or she
Who learns to sculpt in plastic waste;
From stone and wood and such set free
To express in full artistic taste.

A margarine tub or yoghurt cup
Is melted down and squirted out
As rod-stock, right for building up
The parts required for a sculptural bout.

An electric loop, heated red,
In a handle neat and small
Is what we use in a chisel's stead;
No need have we for the sculptor's maul.

All our tools we make ourselves,
The cost is almost nil,
And one who into such efforts delves
Quite quickly gains the needed skill.

Our studio is but three feet square,
We can move it about with ease
And place it here or place it there,
Our special needs to please.

So why not try this happy art,
This art of plastic scrap?
Perhaps you'll find right from the start
That to do it well is just a snap.

In any case, I think you'll find,
As you make the tiny things,
It's a lively task for hand and mind,
And to your days much pleasure brings.